JOUST

Joust is a mechanized variation on the ancient and highly dangerous test of a knight's skills in horsemanship and courage. Each robot takes its chances in a head-on charge against a house robot. In the event that the competing robot is knocked off the jousting platform at the point of impact, wherever the robot comes to rest will be its final position. Should the robot manage to survive this colossal impact, it must attempt to gain as much ground as possible. The game lasts for 30 seconds, after which the worst performer is eliminated.

TUG OF WAR

Tug of War is a game from which there is, quite literally, no escape, since each competing robot is physically chained to one of the infuriated house robots, while between them yawns a deep pit filled with vicious spikes, explosive charges and all-consuming fog. At the end of 30 seconds, if the robot has managed to avoid being dragged in to the 'pit of oblivion', its distance from the pit is measured, with the worst performer – the one nearest the pit – being eliminated.

TECHNICAL MANUAL

ALAN BAKER

BOXTREE

First published in 1999 by Boxtree,
an imprint of Macmillan Publishers Ltd,
25 Eccleston Place, London SW1W 9NF and Basingstoke

Associated companies throughout the world

ISBN 0 7522 1361 X

9 8 7 6 5 4 3 2 1

A CIP catalogue record for this book is available from the British Library

Photographs by Giles Barnard © BBC/Mentorn Barraclough Carey Productions Ltd, except pages 40, 46, 53, 56 and 63 © Mentorn Barraclough Carey Productions Ltd.
Line illustrations by ML Design © Boxtree

Designed by Blackjacks
Colour reproduction by Blackjacks
Printed in England by Bath Press Ltd

mbc
mentorn
barraclough
carey

This book is based on the television series *Robot Wars* produced by TV21 for BBC2.
TV21 is a Mentorn Group Company.
Executive producer: Tom Gutteridge
Producer: Stephen Carsey

CONTENTS

What is Robot Wars? 6

Behind the Scenes 20

The House Robots 38

Building Your Own Robot 68

Robot Wars on the Internet 140

The Robot Wars Club 141

How To Make A Video Diary 142

Index 143

CASSIUS

WHAT IS ROBOT WARS?

Robot Wars is, quite simply, a new and totally revolutionary spectator sport, in which teams of competitors design and build radio-controlled fighting machines, and enter them in mortal combat both with each other and with five terrifying house robots, whose only function is to maim and destroy. In the course of the Wars, the competitors are required to bring a variety of skills to each battle, including engineering and a flare for imaginative and effective design. But, in order to secure victory, they must also display lightning-fast reactions and consummate driving ability.

Robot Wars was originally devised by Marc Thorpe, an American who worked at George Lucas's Industrial Light and Magic special effects company in Hollywood, the company responsible for *Star Wars*. As might be expected with such an original sport, the idea for *Robot Wars* came to Thorpe in a rather unusual way – while he was trying to design a radio-controlled vacuum cleaner to make housework more fun. He remembers, 'I mounted the vacuum on a remote-controlled tank to see whether the idea would work. It didn't vacuum too well, but it got me wondering why not get a tracked vehicle and mount power tools on it?'

While the original *Robot Wars* were fought in America (and are now staged annually in San Francisco), the British version is more complex and visually exciting.

Aside from the individual combat between competitors in the American version, the British Wars also feature a wide variety of rounds, challenges and trials for the guest machines – including a deadly assault course known as the Gauntlet, and a selection of Trials including a robotic version of sumo wrestling! In addition, the competitors' machines must also confront the power, might and hideous weaponry of the house robots – Shunt, Matilda, Dead Metal, Sergeant Bash and the latest addition, Sir Killalot.

The First Wars, a six-episode orgy of twisted metal and spilled engine oil, presided over by the grinning apparition of Jeremy Clarkson, were recorded in November 1997, and hit British television screens at the beginning of 1998. Few of the 4.5 million people who tuned in had ever seen anything like it before. Both critical and public acclaim were swift in coming, with even the harshest critics describing the Wars in the most enthusiastic terms. So successful was the show that a new and expanded series was commissioned – the Second Wars, which contains 18 episodes, and includes a *Making of Robot Wars* documentary.

The British version of *Robot Wars* was developed in March 1995 by the independent production company Mentorn Barraclough Carey, and is the result of intensive research and consultation with university departments, technical experts, radio-control enthusiasts, and designers. In the summer of 1995, pre-production began on a pilot programme, in which four British robots were pitted against three of their mechanized American cousins. The 1995 British *Robot Wars* Championships were held in a warehouse in west London, before an invited audience of press and public. Although the American robots – La Machine, Thor, and the Master – thoroughly vanquished their British opponents, everyone was so impressed by the energy and excitement of the Championships that negotiations for the development of the First Wars began shortly afterwards.

In January 1997, a small and dedicated production team began the process of recruiting the roboteers who would do battle, enlisting the aid of schools, universities, special-effects experts, and engineers, all of whom responded enthusiastically to this strange but hugely exciting and challenging idea. In November of the same year, recording began in a large warehouse complex in east London's Docklands, with Jeremy Clarkson in the role of Commander-in-Chief, and Philippa Forrester as Pit Reporter, conducting on-the-spot interviews with the roboteers as they repaired and fine-tuned their machines, readying them for each new encounter.

An intrepid roboteer works against the clock to get his machine up and running.

The Mule and Demon confront each other.

The First Wars featured 36 teams of 155 roboteers, doing battle in a set created largely (and rather appropriately) from pieces of scrap found in an aircraft graveyard where Tornado fighter planes meet their final destiny. The excitement and fun, fast-paced action, flying metal, and outrageous robot design ensured the success of *Robot Wars*.

While the competitive structure of the Second Wars is essentially the same as the First, many improvements have been made – in other words, *more destruction, more violence*, and *more fights to the death*! There are also many more robots prepared to engage in combat, both with each other and the foul-tempered house robots. While the First Wars saw the participation of 36 robots, the Second Wars feature no fewer than 72 machines, battling through 15 episodes of competition and even more robots feature in special programmes. The proud victors of each episode eventually meet each other and the house robots in two semi-finals and the grand final, while the vanquished return in ignominy to the scrap heaps that spawned them. But that's not all. Not content with all this action, the producer has also included several head-to-head grudge matches between the mightiest and strongest robots, a documentary which shows how *Robot Wars* is made, and a series of weight-specific contests for the lighter-weight machines.

Though the action predominantly involves Heavyweight robots, there are four weight categories recognised in the world of *Robot Wars*: Featherweight, up to 11.4 kilograms; Lightweight, 11.4–22.7 kilograms; Middleweight, 22.7– 45.4 kilograms; and Heavyweight, 45.4–79.4 kilograms.

The aim of each team is to take their robot through the trials and tribulations of the Gauntlet and the Trials (with the worst performer being eliminated from the competition at each stage), and on to the Arena, where the victors meet each other in mortal combat.

THE GAUNTLET

This is a deadly assault course designed to test the speed, strength, and manoeuvrability of the robots, as well as the navigational skills of the drivers on their way to the End Zone. In the First Wars, the Gauntlet's starting point was an enclosed revolving platform. Any delay in starting meant the robot would be facing in the wrong direction, and would be at an immediate disadvantage against the clock. Good timing at the start was of the essence. Once into the Arena, the robot had a choice of three initial obstacles to negotiate: including the Maze, which required extreme manoeuvrability (any machine with a turning circle much above zero would have problems); an apparently dead space, occupied by one of the house robots; and a see-saw ramp, a deceptively innocuous looking feature that nevertheless gave several robots with low ground clearance serious problems. Once through these initial obstacles, the competitors' robots had to avoid a set of vicious corkscrew spikes which periodically thrust into the Arena, another set of spikes rising suddenly and unexpectedly up from the floor, assorted inactive features such as giant lorry suspension springs, a spiked pendulum ball capable of knocking the lighter robots sideways and costing them extra time, and two more house robots, both of whom would do everything in their power to stop the competitors' machines from reaching the End Zone.

Dreadnaut falls foul of Sir Killalot, Shunt and Matilda.

In the Second Wars, the Gauntlet is even more challenging, though the rotating starting block has been discarded in favour of an open-ended steel cage. There are three routes to choose from which change from week to week. One course requires the competitors' robots to avoid rising spikes in the floor, then to batter their way through a brick wall, and negotiate the see-saw ramp before having to deal with one of the house robots. Or they may have to face a ram-rig, in which unluckier robots will be pushed inexorably towards a bank of circular saws, while those who manage to get out the other side in one piece will find themselves having to avoid the Flame Pit, like some hideous, man-made volcano. One of the house robots will be watching this section of the Gauntlet. Another alternative in this newly designed Gauntlet is an apparently clear run towards

The Gauntlet awaits its first victims.

the End Zone, with only two pits to avoid. However, between these pits stands yet another house robot, its mission to prevent the competitors' robot from completing the course.

In both the First and Second Wars, the Gauntlet also features one of the most effective obstacles in the whole competition – the tank traps – which instantly immobilize any machine unfortunate enough to drive (or be pushed) on to them.

By the end of the Gauntlet, at least one robot will be eliminated through damage, destruction, or by simply not managing to cross the finish line. In the familiar event of more than one failure to reach the End Zone, whichever robot is furthest from the finish line will be politely asked to leave. If all the robots manage to complete the course, the one with the slowest time will be eliminated.

THE TRIALS

The Trials await those robots that have emerged intact from the Gauntlet. After a quick pit-stop for any repairs and maintenance needed, the robots will be subjected to a series of extreme challenges and tests, each of which is designed to cause as much damage and discomfort as possible. This part of *Robot Wars* can best be described with two words: *violence* and *destruction*.

The Trials in the Second Wars feature six games, with a seventh reserved for the semi-finals:

SUMO

Sumo is played on a raised circular plinth, 4.8 metres in diameter. Each of the surviving robots goes head-to-head against one of the implacable house robots. The game has a time limit of 30 seconds, during which each robot must attempt to push the other over the edge of the plinth. At the end of the conflict, whichever robot lasted the least amount of time before being shunted over the edge is eliminated.

KING OF THE CASTLE

King of the Castle, like Sumo, lasts a maximum of 30 seconds. Each robot takes it in turn to hold the high ground against a series of vicious assaults from one or more of the house robots, against the clock. Whoever is removed from the platform in the least amount of time will be eliminated.

SKITTLES

Skittles is not quite as tame as its name suggests, since each robot must attempt to knock down as many barrels as possible in the time allowed, while simultaneously battling with the metal-hungry house robots, out, as always, to cause maximum destruction.

SOCCER

Soccer is another variation on a more familiar sport, in which all the competing robots attempt to outscore each other. When a robot scores, the game is stopped, the scorer is removed from the field and proceeds to the next round, the ball is repositioned and the game then resumes with the remaining robots. At the end of the game, whichever robot is left on the field is shown the red card. This game is a particular test of competitors' driving skills, not least because the goal is defended by the house robots.

PINBALL

Pinball – a fiendish game – is kept in reserve for use in the semi-finals. In this game, each robot has to score as many points as possible by completing certain challenges, such as navigating obstacles, knocking down targets, taking on the house robots and so on. The harder the challenge, the more points will be at stake. Whoever has the lowest score when the clock hits zero will be eliminated.

THE ARENA

The Arena is the place where only the toughest meet: the four robots that have come through all the previous encounters with each other, the various obstacle courses, and survived the house robots, intact and still functioning. The battles taking place within the Arena climax in two semi-finals and a grand final.

The Arena is nothing less than an all-out combat zone, featuring a wide variety of hazards (both active and fixed), traps, and house robots guaranteed to be particularly annoyed with the upstarts that have managed to get this far. The object for the competitors is to destroy or immobilize each other's machines, while avoiding the unpleasant attentions of the house robots who defend from the perimeter zone until, by the end of each heat, there is one winner, who then goes through to compete in one of the semi-finals of the Second Wars.

THE CREW

Robot Wars relies for its incredible excitement on the talents and commitment of a large number of people. Aside from the vital roles played by Craig Charles – who replaces Jeremy Clarkson as Commander-in-Chief in the Second Wars, Philippa Forrester as Pit Reporter, and Commentator Jonathan Pearce – an enormous amount of behind-the-scenes work is required to make the Wars the huge success they have been. More than 150 people work on the show, from production staff to audience ushers, set designers to camera operators and sound engineers to security personnel.

Preparation for the Second Wars began in April 1998, and the Wars were recorded in just 10 days at the Royal Victoria Dock in east London.

MACE

PANIC
TTACK

BEHIND THE SCENES

A programme such as *Robot Wars* is the result of the expertise, inspiration, and sheer hard work of many people, in this case more than 150. Tom Gutteridge, Chief Executive of production company, Mentorn, had the idea for adapting the American spectator sport into a television show after seeing a tape of the American version. He decided to invest money into developing the format and bought the rights to the *Robot Wars* name. He then engaged Stephen Carsey, the producer of British children's show *Scratchy & Co*, to help Mentorn devise the concept. For Tom and Stephen *Robot Wars* is the culmination of three years of intensive development work.

Four weeks before shooting on the show began, a group of prospective roboteers gathered in a Docklands warehouse in east London to take part in what surely must be among the strangest auditions in the history of showbusiness. Their machines were examined by the organizers, who made copious notes on

Judge Adam Harper once caused a problem by wearing a bright orange tie on the set. Since bright colours and certain patterns can cause a strobing effect on television, the crew had to check whether Adam's tie would strobe.

For Peter Duncanson and his Spin Doctor team, the best memory of *Robot Wars* is the camaraderie between teams. He tells of one occasion in particular, when team-mate Martin Griffin put his foot on Spin Doctor and asked for full power, to see how powerful the robot was. Unfortunately, the speed controller promptly burned out, but two other teams stepped in and donated replacements, thus saving Spin Doctor from an ignominious withdrawal from the Wars.

design, propulsion, and weapons elements, while the radio-control systems were submitted to Transmitter Control for approval. This was one of the most important aspects of the production since it was at this point that the machines were checked to make sure that they conformed to all of the health and safety regulations. Safety, of course, is paramount, for the studio audience and production crew as well as for the roboteers themselves. So, for example, weapons such as untethered missiles, explosive and corrosive substances are not allowed. (After all, the point is to make robots suffer – not humans!) In addition, it was important to make absolutely certain that the teams of roboteers knew exactly what would be expected of them and their machines, in terms of commitment, competence, and stability, under the sometimes enormous pressure to which they would be subjected during the filming of the Wars themselves.

In one of the more bizarre pre-production moments, Stephen also used the audition as an opportunity to test one of the challenges that had been devised for *Robot Wars*. While the roboteers were busy answering various technical questions on their machines (and in some cases trying to estimate how much they had actually spent on them – not an easy task, considering how much material is found or donated), Stephen, together with a colleague, drew a life-size plan of the Maze (one of the obstacles in the Gauntlet) on the floor of the warehouse in which the audition was taking place. He then asked the roboteers to guide their machines through it, while imagining that the chalk lines were actually walls. The drivers showed remarkable skill, and the Maze was included in the final selection of challenges.

For Stephen, watching *Robot Wars* come to life, from an initial idea, through development and production, to a fully fledged television show was a hugely exciting and rewarding experience. He is also keen to acknowledge that *Robot Wars* is only as good as the competitors and robots taking part. If the machines had broken down, or had not worked at all, then neither would the show. Fortunately, no such disasters befell *Robot Wars*; the robots worked well, and the competing teams themselves pulled together with each other and the production crew to create the sense of a shared endeavour, which resulted in a show that works brilliantly.

For Peter Gibson, the worst experience he had in *Robot Wars* was during the initial parade, when the radio controller suddenly failed, and Wheelosaurus meandered aimlessly off into the smoke.

For the Napalm team's Claire Greenaway and Becci Glenn, getting knocked out early in the First Wars was something of a blessing in disguise, since afterwards they were given jobs to do in the backstage area. During this time, they learned a lot about the design and construction of the house robots, and this experience not only inspired them to continue their efforts in the Second Wars, but also gave them some ideas to include in their design for Napalm.

BUILDING THE ARENA

One of the three main elements in *Robot Wars* is the Arena – a noisy, smoky, post-apocalyptic space in which all the action takes place. Some 30 people were involved in its construction, an operation which, due to the pressures of the shooting schedule, had to be completed in three days – a somewhat nerve-wracking but ultimately rewarding experience for the designers and builders. The first priority was to get the stage itself into place, followed by the construction of the scenery and lighting around it. The floor panels were raised above ground level to allow room for the machinery that operated the active components of the Arena, such as the Flame Pit and the vicious spikes that rise up through the floor. The majority of the scenery was constructed from large pieces of metal retrieved from scrap yards. Scrap metal was used for two reasons. Firstly, because it looks good with a coat of paint, and secondly because it is cheap. The addition of neon lighting (always telegenic) completed the grungy, industrial look of the Arena.

For production designer Julian Fullalove, the call to the Wars came as a result of some friends whose company had been approached to design and build the house robots. Though that fell through, they told him about Mentorn's plans for *Robot Wars* and that they were looking for designers to submit ideas. Julian, who had previously worked on the British children's television series *Teletubbies* and the spy drama *BUGS*, was among six top designers who submitted ideas to Tom Gutteridge and Stephen Carsey. They chose Julian's after seeing a conceptual model of the *Robot Wars* stage set. The model took two days to design and one day to build, and was

The Steeples brothers, Oliver and Ben, make adjustments to their robot, Griffon.

based on the concept of an amphitheatre – with the seated audience split into two banks on either side of the central Arena, well away from any danger (which could raise its ugly head quite suddenly and unexpectedly).

Working constantly against their unavoidable budget constraints, Julian and his team built the set around a stage area raised 1 metre above the floor of the Docklands warehouse in which the Wars were to be filmed. There were three main reasons for raising the floor. Firstly, as mentioned above, space was needed to install the equipment for the active elements of the Arena and the pits backstage; secondly, a raised stage

made it much easier to manipulate the cameras, and provided them with a greater variety of angles and positions from which to shoot; and thirdly, the floor of the warehouse was somewhat uneven, and as such was pretty useless for the production's intentions. This problem was overcome with the use of 2.5-by-1.2-metre rostra, supporting a tough and durable plywood surface. Each of the rostra had screw-out feet, which could be adjusted to provide a completely flat surface on which to conduct the Wars. In spite of its toughness, the plywood surface took an enormous amount of punishment from the robots, and quickly acquired a battered and beaten look, which actually made the Arena look more like an authentic mechanical battlefield.

For the scenery components, Julian and his team went to Hanningfield Metals in east London, where they spent a good deal of time rummaging around the scrap yard, which contained a wide variety of aircraft panels, nosecones, missile components, and other assorted military cast-offs. They then hired an enormous skip, filled it with £2000-worth of dead and twisted metal and brought it back to the warehouse, where they then proceeded to attach it to the mesh screens surrounding the central Arena.

Julian had originally wanted to install in-floor lighting in the Arena, but found that the robots quickly destroyed the wiring, leaving live wires lying around the stage – obviously unacceptable from the point of view of safety (particularly the safety of the camera operators working on the Arena floor), and so, unfortunately, the idea had to be abandoned. However, no in-floor lighting meant that more floor spikes could be added, as well as a much more threatening swinging ball – the one in the First Wars weighed about 22 kilograms, and failed to do

The main set of *Robot Wars*, including the entrance to the Arena and the contestant control booths.

David Barker of the Sting team remembered that he had booked a holiday in Wales, during which the team were struggling to replace the robot's sheared drive gear. At one point, in a rather surreal moment, David found himself talking on a mobile phone to team-mate Ian Pritchard about getting hold of a replacement gear for their robot ... from the top of Mount Snowdon!

For the Razer team, the best memory of *Robot Wars* was making fun of Pit Reporter Philippa Forrester. One of the time-saving techniques used in television interviews is for the interviewer to be filmed doing nothing but nodding, as if listening to the answer to a question; this is then edited into the interview footage at a later stage. Ian says that he and Simon tried desperately to make Philippa laugh while she was doing this, but without any success whatsoever. In spite of their best efforts, Philippa remained totally straight-faced throughout. Ian and Simon could not help but be impressed by her professionalism.

significant damage to any of the robots; the one used in the Second Wars weighs over 500 kilograms, and must now be avoided at all costs.

Julian's worst memory of working on *Robot Wars* was watching Sir Killalot take out his frustration on the set, and so on all the hard work the team had put in. Whenever he missed the opportunity of crushing a competitor's robot to death, Sir Killalot's foul temper apparently got the better of him, and he would wrench chunks out of the set's scaffolding and handrails with his powerful cutting arm. As a result, Julian found himself repairing the Arena a little more often than he had anticipated.

Julian encountered another problem when trying to retrieve robots from the one of the pits, which had been added for the Second Wars. On a number of occasions, when a robot fell in upside-down, its master power switch became unreachable, and those charged with

Sir Killalot goes to work.

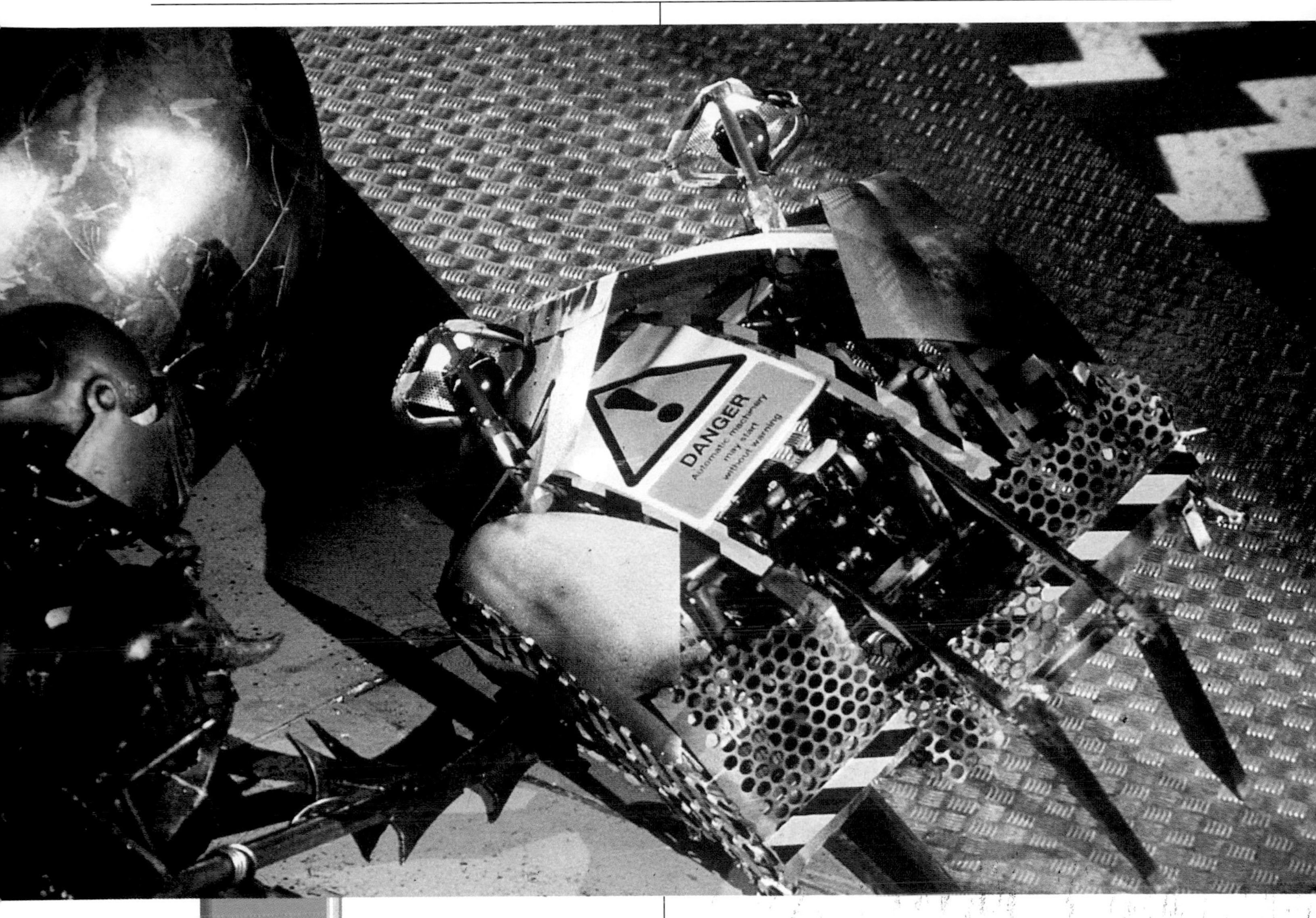

Sir Killalot takes an unhealthy dislike to Napalm.

the responsibility of retrieving it found themselves constantly dodging flailing lifting arms and uncontrollably spinning cutting discs.

The construction manager in the Second Wars was Paul Duff, who has a background in film and television. Paul noted that the stage in the First Wars took so much damage from the robots that it had to be entirely upgraded and reinforced. The robots taking part in the Second Wars, he said, were much stronger, more powerful and better built than those in the First Wars, requiring reinforcements on all the barriers surrounding the Arena, as well as the floor itself, which is composed of 2.5-metre tiles corresponding to the units underneath. These tiles had the advantage of being easily removable, both to allow access for maintenance, and to create the pits into which the competitors' robots could be slung (often by Sir Killalot).

Paul agreed with Julian that the sight of Sir Killalot venting his spleen on the stage set was somewhat alarming and disheartening. He also described how robots consigned to the Flame Pit had to be removed quickly but very carefully, as their engines and weapons were often still active. Paul likened them to giant insects buzzing angrily on their backs.

Despite the fact that the 12- to 14-hour shoots and six set changes each day meant a lot of hard work, his involvement in *Robot Wars* was enormous fun, and was a hugely novel and rewarding experience.

LIGHTING

Pit Reporter Philippa Forrester surveys the battlefield.

Lighting is a particularly important aspect of any television production, and *Robot Wars* is no exception. Lighting director Bernie Davis watched the First Wars, and was very impressed. When the programme's director, Stuart McDonald, with whom he had worked on other shows, suggested he consider working on *Robot Wars*, Bernie met with producer Stephen Carsey. According to Bernie, Stephen was looking for a very definite look to the Wars. One thing he didn't want was for it to look like light entertainment – a kind of 'robot disco'. Stephen was aiming to give the impression of a dark, moody, threatening subterranean space, as if the roboteers had discovered the long-abandoned relic of a sinister mechanistic culture – the perfect environment in which to fight their Wars.

To achieve this effect, it was important for the lighting to accentuate the sharp angles and hard edges of the set. This was achieved by using a complex lighting system consisting of eight lighting chassis suspended over the Arena, in addition to 12 cyberlights – moving lights whose beams are projected on to mirrors, thus generating constantly changing patterns. There were also 20 varilites, which were also capable of moving and painting colour washes – patterns of light that could be varied in their intensity, from soft circles to sharply defined pinpoints. The initial lighting design process took four days of meetings and drawings, before the final design was settled on.

Bernie's one particular problem, rather ironically, was with smoke in the Arena. They ended up generating a certain amount themselves in order to accentuate the various beams of light playing over the set; but, the robots also produce their own smoke, and sometimes there was so much that it almost obscured the action.

WRITING

For the Second Wars, producer Stephen Carsey recruited Nathan Cockerill as writer. The original Commander-in-Chief, Jeremy Clarkson, wrote his own scripts for the First Wars together with producer Stephen Carsey. However, with 18 shows in the Second Wars time was an ever-present factor, and Stephen decided to employ a professional writer. Nathan had already worked with Craig Charles on the British television early morning show the *Big Breakfast*, and they worked so well together that Nathan seemed the logical choice to work on *Robot Wars*.

Nathan says that he had a definite brief for writing the Second Wars. It was very important to Stephen that the audience were not patronized by the implication that *Robot Wars* was actually 'happening', in other words, that the action was taking place on some other planet or at some other time, as had been the conceit with certain other television shows. In spite of its post-apocalyptic, science fiction overtones, *Robot Wars* is not science fiction, it is a genuine – albeit highly unusual – sport.

The Second Wars took Nathan three days to write. This might seem a surprisingly short time, until one remembers that *Robot Wars* is centred around the action in the Arena, and the time spent by Craig Charles actually talking to camera is measured in seconds. For her part, Philippa Forrester always ad-libbed while guiding the television audience through the pit area and during the brief interviews with the competitors, and Nathan was impressed with the effortless way in which she handled her part and left her to it.

Commander-in-Chief Craig Charles receives some maintenance of his own.

Philippa Forrester interviews one of the teams in the pits area.

Post-battle celebrations and commiserations.

THE JUDGES

Another important element in *Robot Wars* is the judging. There are four basic criteria for the Judges to consider: damage, aggression, control and style, for each of which they can award a total of four points. The points awarded for damage are then multiplied by four, aggression by three, control by two and style by one. The scores are then added up to give the final score.

The three Judges featured in both the First and Second Wars are Adam Harper, Eric Dickenson, and Professor Noel Sharkey, all of whom are uniquely qualified to meet the responsibility of judging the competition. Adam Harper holds the land speed record for an electric vehicle, which he achieved in a modified Sinclair C5 reaching 241 kph (150 mph). Adam was originally a technical advisor to the series; however, just

Adam Harper remembers how the Judges had a habit of annoying the floor manager. The Judges' booth is situated quite high up on the set, and gets extremely hot due to the lights and the activity of the robots (particularly Sergeant Bash and his flame-thrower). Since hot air rises, the Judges ended up sweltering in their booth, and kept dashing off to the canteen for iced water every chance they got. For obvious reasons, this gave the floor manager something of a headache, and he could often be seen stalking into the canteen and telling them to get back to their booth.

days before shooting was scheduled to begin, he was asked if he wanted to be a Judge.

It was through his son's interest in racing model cars that Eric Dickenson saw a video of the US *Robot Wars*, and became interested. After consulting several Internet websites, he built his own robot and took part in the US Wars, where he met Stephen Carsey, who was filming there at the time. Eric subsequently entered the UK Wars but had to withdraw. Nevertheless, Eric had considerable robot building experience, and so Stephen asked him if he would like to be a Judge. Eric took part in a total of four US competitions – as well as a competition between two teams of four robots, in which his 11-kilogram robot knocked the drive chain off a 75-kilogram French entry, which pleased him no end.

Judge Adam Harper says that there were a lot of times when the Judges had very little to do (such is the nature of shooting a television programme), and at one stage, in order to relieve their boredom, he and Professor Sharkey decided to have a chair race in the Judges' booth. Adam says that he won, although this is disputed by the professor.

Professor Noel Sharkey is Head of the Department of Computer Science at Sheffield University in the UK and Chairman of the Artificial Intelligence and Neural Computers Committees of the Institute of Electrical Engineers, and thus was another excellent choice for the judging in *Robot Wars*. Professor Sharkey is particularly interested in the controversy over whether the robots on *Robot Wars* are actually robots, as opposed to glorified radio controlled cars. He believes that the correct description is 'tele-robots', or remotely-controlled robots. (For those still reluctant to accept that these machines really are robots, one of the dictionary definitions of the word 'robot' is, 'any machine or device that works automatically or by remote control').

Craig Charles with the *Robot Wars* Judges (from left to right) Adam Harper, Professor Noel Sharkey and Eric Dickenson.

However, Professor Sharkey would very much like to see more autonomous machines in *Robot Wars* – machines capable of reacting to external stimuli with no intervention from the roboteer and this is something the production team are keen to encourage. There are a number of actions that can be programmed into a robot. For instance, a human operator may trigger an attack or escape sequence, which the robot will then follow automatically. There is also the additional possibility of using various sensing devices, such as ultrasound and infrared sensors, to detect obstacles and automatically avoid them. A much more ambitious idea, however, would be to include a program that enabled a robot to attack other machines automatically on sight. Of course, this would mean installing cameras on the robot, and teaching it how to recognize enemy robots. For this, you would need a machine equipped with what are known as genetic algorithms – programs that actually evolve and improve, or neural networks, that are capable of learning. The raw materials required for such a robot are quite inexpensive. However, Professor Sharkey estimates that the research required to arrive at a working model would put the price tag around the £50,000 mark – not an attractive proposition with Sir Killalot waiting in the wings to destroy it.

With regard to judging the Wars, Professor Sharkey feels that it is sometimes difficult to award the right scores, particularly in terms of aggression. Ultimately, the Judges found themselves having to differentiate between aggression and *force* of aggression. For example, a small robot could be flattened by a single blow from a large machine; and yet it could have made numerous attacks, to the very best of its ability, on its opponent before finally succumbing. The Judges had to take both these aspects of the battle into account when coming to their decisions.

For many who saw it the best moment during the filming of *Robot Wars* was Ben Symons' demonstration of Broot's weapon, the 'untz' (an 'untz' being a punch usually delivered in the school playground) which came after Pit Reporter Philippa Forrester made the mistake of asking him what an 'untz' was. He replied to her question by giving her a quick jab in the ribs, mimicking the action of the robot's spring-loaded ram, and which caught the intrepid Philippa by surprise.

The Sentinal, a defensive obstacle that stands guard during the Gaunlet.

For Neil Lambeth and John Ebdon, the best moment in *Robot Wars* was actually managing to drive Elvis out in the initial parade, since the robot was very nearly as wide as the door through which he was supposed to go. The clearance was only 2 centimetres on each side.

THE AUDIENCE

One of the things that makes *Robot Wars* such an exciting event is the presence of the extremely enthusiastic audience, many of whom can be seen bashing their fists against the perspex shielding around the Arena. It is easy to forget how much planning and effort goes into looking after a large studio audience. However, audience gatherer Celia Nicholas knows exactly what is involved in making sure all goes well with the 400 people (in two daily sessions of 200 each) who come to experience *Robot Wars* live. One of the things Celia must watch out for is people standing up during the events, which is absolutely forbidden for the audience's own protection. *Robot Wars* is a dangerous sport (as indicated by the bullet-proof perspex lining the Arena). For the same reasons Celia must also ask people not to leave during events. She also has the unpleasant task of making sure that no one under the age of eight is allowed in – loud noises (which *Robot Wars* contains in abundance) can be extremely upsetting to some young children. Celia remembers the sad but (it has to be said) rather amusing story of the young boy who thought that the voice of director Stuart McDonald, which comes through over a loud-speaker in the Arena, was the disembodied voice of a demon! The poor lad had to be taken out to watch the remainder of the session on the monitor in the audience holding area.

As already mentioned, health and safety is of paramount importance in *Robot Wars*, as in any television event, and Celia and her team were constantly on the lookout for any problems, staying in constant contact with each other via walkie-talkie. A day's shooting lasts from seven in the morning until seven in the evening. It's a long day. Thankfully most of the incidents they had to deal with were more irritating than dangerous. On one occasion, after a night of particularly heavy rain, all the portable toilets malfunctioned. Celia and her colleagues ended up having to usher the members of the audience through the pits area (normally off limits) to the toilet facilities in the robot loading docks.

GRAPHICS

One of the most striking visual aspects of *Robot Wars* is the highly inventive use of computer graphics, evident right from the opening title sequence. The title graphics were designed and produced by Matt Clark at Mind's Eye, a company which also produces computer games for PC and PlayStation. Matt had quite an open brief from Stephen Carsey. This freedom allowed Matt and his team to experiment with a fairly abstract set of images. They began with a set of storyboards – black-and-white hand drawings of all the scenes they envisaged for the opening sequence (a technique also used in conventional film-making). The next step was to convert the drawings to a three-dimensional computer version. At this stage, all the images were frameworks of lines, with no other detail. The background environment was then added, along with the various metallic textures of the objects, which included robots, cogwheels, weapons, and so on. Finally, the artificial lighting was added, along with pieces of fine detail. This was achieved using SoftImage and Kinetix 3D Studio Max software. The entire process took about three months.

The graphics team had been aiming for a certain feel, reminiscent of other similar British television shows like *Gladiators* and the *Crystal Maze*, but with an additional dark, industrial feel, using the various mechanical components coming together to form the robots. The initial design, however, showed the robots ripping each other apart, which was considered a little *too* dark, especially in view of the younger audience who

The worst moment for the Haardvark team came when they tired to test the circuitry, and it immediately blew up. In fact, during the Wars, the team checked out of their hotel every morning, assuming that they would be knocked out that day! With a certain amount of embarrassment mixed with relief, they had to ask for their rooms back several times.

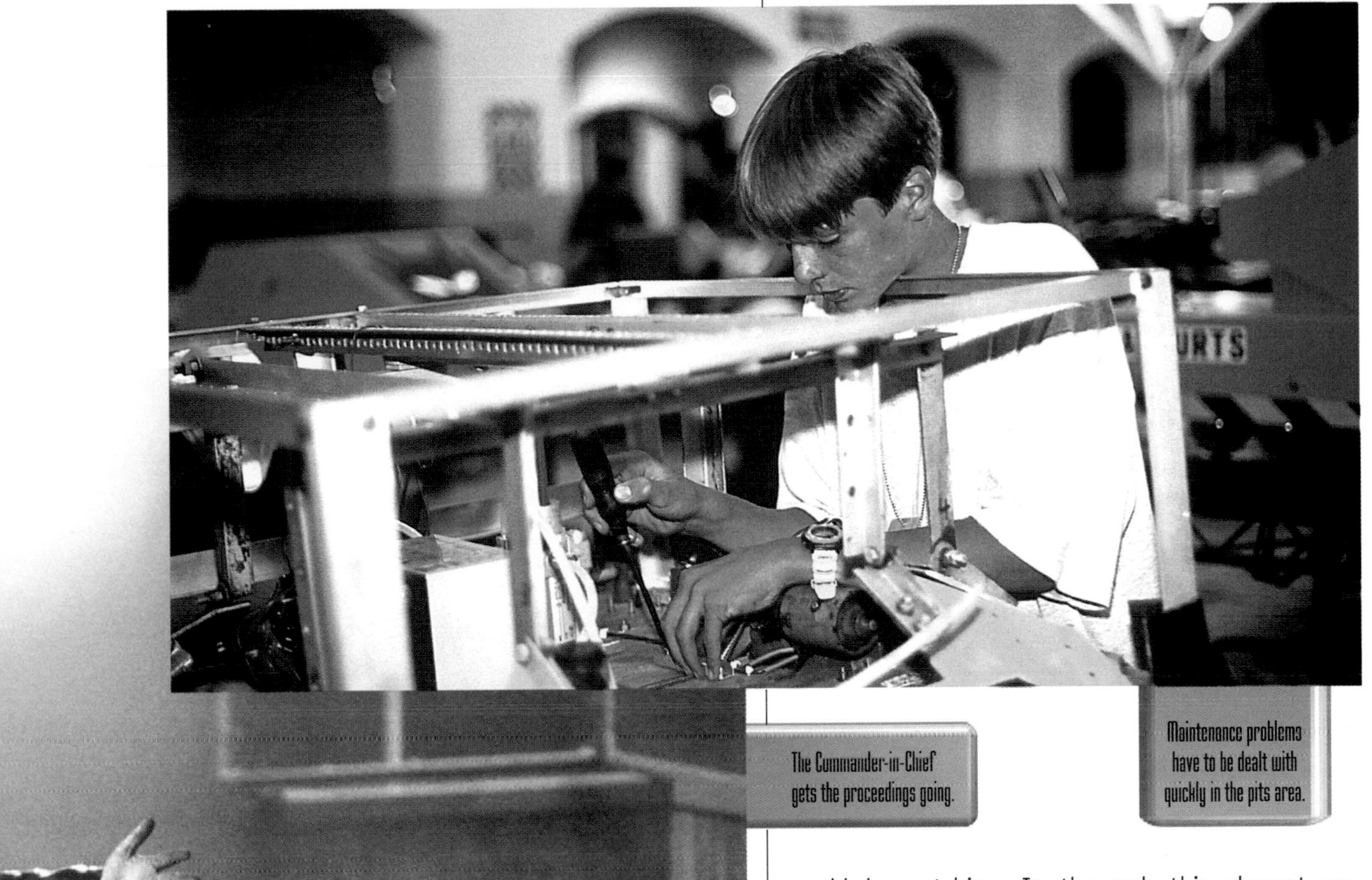

The Commander-in-Chief gets the proceedings going.

Maintenance problems have to be dealt with quickly in the pits area.

would be watching. In the end, this element was reduced to an arm pulling a single cog out of position.

The graphics for the competitors' robots' specifications in the Second Wars were designed by Craig Howarth at 4:2:2 in London. There were no particular guidelines for Craig to follow, aside from maintaining a functional look, which was consistent with the First Wars, using the *Robot Wars* house-style lettering. The specification panels and results boards featuring parts of the competitors' robots were produced using a device called a 'cap-gen' or caption generator. Craig says that the same effect could have been achieved with a computer package such as PhotoShop, although the caption generator is much faster (a definite advantage, considering the large number of robots that had to be dealt with). In fact, according to Craig the only problem they had was the huge amount of graphics that had to be produced to cover the 72 competing robots. The machines themselves were photographed before filming began, and their images combined with the background of the specification panels via the caption generator.

THE COMMENTATOR

The distinctive voice and infectious enthusiasm of Commentator Jonathan Pearce greatly enhances the excitement of *Robot Wars*. Jonathan himself finds the idea of *Robot Wars* exciting and intriguing, since it works extremely well as a theatrical event as well as a legitimate sport. However, he doubts that it will replace sports in which human beings compete directly with each other. But, of course, as technology improves and the range of entertainment media continues to grow, there will be a greater and greater demand for both traditional sports and their unusual and innovative counterparts. It must be remembered that the UK *Robot Wars* are only two seasons old, and it is likely that they will continue to grow in popularity, until they have taken their rightful place alongside more traditional sports.

THE PRESENTERS

Robot Wars is fronted by two faces familiar to the British television audience: Philippa Forrester as Pit Reporter and Craig Charles as Commander-in-Chief. Philippa was a natural choice for Mentorn because of her previous experience on *Tomorrow's World*, the popular BBC science show. With her interest in technology she was keen to get involved in what she considered a bizarre but exciting idea. Although Craig was not the Commander-in-Chief on the First Wars he was a big fan of *Robot Wars* and so was very pleased to be offered the role for the Second Wars, as was his son who thought Craig was 'seriously cool'. However, it seemed that filming of the comedy series *Red Dwarf* was going to clash with *Robot Wars* and so initially he had to turn Mentorn down. But Stephen Carsey felt that Craig was exactly the right man for the job and so just days before filming he called Craig's agent one last time. The filming for *Red Dwarf* had been moved and Craig was available after all.

For both Craig and Philippa filming *Robot Wars* is both enjoyable and rewarding not least because it is creating enthusiasm for technology and engineering. For them, however, the stars of the show are the 'mad but clever' competitors without whom *Robot Wars* would not be the success that it is.

For David Crosby, Claire Greenaway, and Becci Glenn, the most memorable moment of *Robot Wars* happened after the competition. Having just asked for, and received, autographs from Commander-in-Chief Craig Charles and Pit Reporter Philippa Forrester, the Napalm team were then surrounded by children asking for *their* autographs.

Craig gets as close as he dares to Dead Metal.

Sir Killalot, the newest and deadliest of the house robots, proceeds to do battle with Piece de Resistance.

THE HOUSE ROBOTS

The most memorable characters on *Robot Wars* are, however, the house robots. Tom Gutteridge and Stephen Casey felt that they wanted the British version to have more violent action than in the American spectator sport. In particular Tom wanted to include fire and explosions and all the other additions that, for safety reasons, were not allowed in the American version. They decided that the only way to include these was to create their own robots which could carry more dangerous weapons under the control of a qualified team. The idea for house robots developed becasue they wanted to give the viewers something to identify with each week. The house robots also gave the competitors additional hazards, rather than simply fighting each other as they do in the American sport.

Mentorn invested huge amounts of money in developing the robots, each of which had to have a definable personality, and, most importantly, be virtually invincible. Having made some conceptual drawings, Mentorn put the job of building them out to tender and the BBC Visual Effects Department (VED) won the commission.

THE HOUSE ROBOTS

There are many ways for a robot to die. As we have seen, the hazards and obstacles encountered throughout *Robot Wars* can knock out a machine in an instant, transforming it from a gleaming, battle-ready device, the pride and joy of its designers and builders ... to a twisted and useless lump of tortured, flame-grilled metal. One of the most familiar and depressing sights in the pits is that of roboteers shaking their heads in despair over the battered bodywork and fried innards of a hapless robot, their dreams of glory mutating into nightmares of humiliation and defeat. More often than not, the harbingers of this ignominy are the terrifying house robots, belligerent masters of the theatre of war.

Each of these dreadful machines has its own characteristics and weaponry. What unites them is implacable ferocity, and an obsessive hatred of the home-made robots that have the temerity to challenge their superiority on the battlefield.

No one is absolutely sure about the origins of the house robots. As we shall see, there are many theories, opinions, assumptions, rumours and wild guesses ... but few hard facts. Some experts have suggested that the house robots have been sent back from the future for some nefarious purpose, while others maintain that they are the result of military experiments that have gone disastrously wrong. As if this were not weird enough, representatives of the BBC's Visual Effects Department (VED) have recently stepped into the controversy to claim that they built the horrific machines. As we shall see, there are probably elements of truth in all these claims, although sorting fact from fantasy may prove easier said than done.

What little information we do have on the house robots is presented here for the first time.

SHUNT

Looking like a cross between a snowplough, a bulldozer and a steam-driven tank, Shunt presents a depressing sight to any competing robot unlucky enough to encounter him. His extreme power and very low ground clearance mean that he is virtually impossible to turn over – or, for that matter, even to move. Shunt goes where *he* wants to go, and there is precious little anyone else can do about it.

Shunt's origin is as mysterious as that of the other house robots, although some attempts at analysis have been made on all of them (with the investigating scientists demanding – and receiving – astronomical levels of danger money). However, in *Robot Wars*, hard scientific research goes hand-in-hand with intriguing rumour and speculation, and Shunt is no exception. One of the most interesting (although unsubstantiated) reports was discovered on an Internet website that has since mysteriously disappeared. The report states that Shunt was originally a remote-controlled drone for use in nuclear reactors. Such drones are used to clean and maintain those parts of the reactor that are too 'hot' for humans, even with protective radiation-proof clothing.

According to this strange and sinister report, about five years ago the drone now known as Shunt was busy conducting routine maintenance on an experimental reactor somewhere in the former Soviet Union, when a sudden power surge occurred, threatening to destroy the entire facility. The engineers fought frantically to bring the reactor back under control, and when eventually they succeeded, they assumed that the maintenance drone had been completely vaporized. The reactor was shut down, and two more drones, equipped with video cameras, were sent in to discover what had happened.

What they saw sent a wave of terror through the reactor control room, some scientists fainted on the spot, while others fled in panic from the facility and were never seen again. The first maintenance drone had *not* been destroyed. Through some unimaginable atomic process, it had mutated into a horned monstrosity, a metallic demon that immediately attacked the other drones, utterly destroying them in seconds. It had become ... Shunt!

All attempts to capture the mutant drone met with failure. It seems that the nuclear power surge had created a thinking brain out of the mindless internal control circuitry, with the result that Shunt became an intelligent electronic entity, capable of strategic planning, and able to elude his human pursuers. He easily escaped from the confines of the experimental nuclear facility, and was not seen again ... until *Robot Wars*.

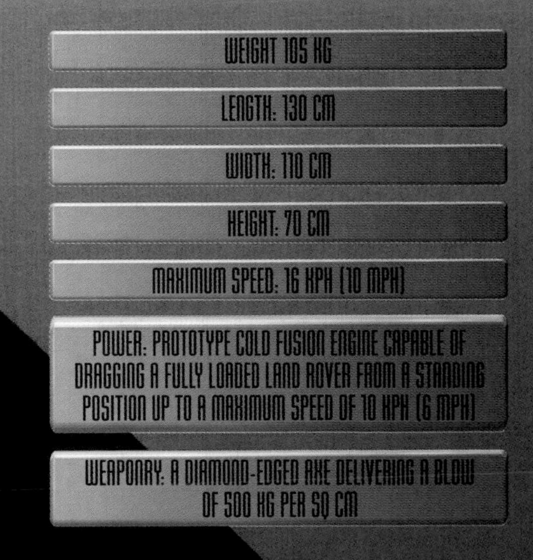

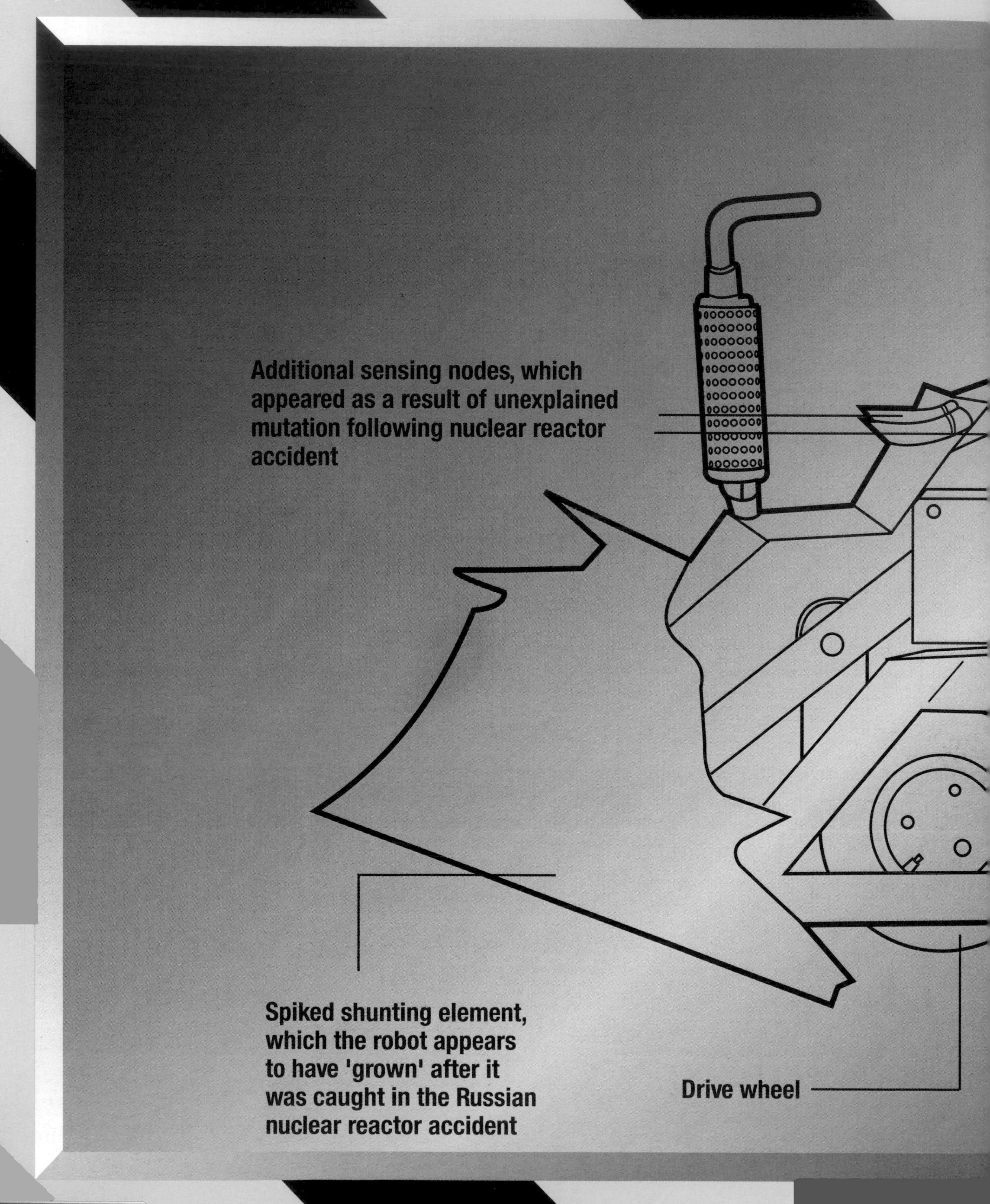
Additional sensing nodes, which appeared as a result of unexplained mutation following nuclear reactor accident
Spiked shunting element, which the robot appears to have 'grown' after it was caught in the Russian nuclear reactor accident
Drive wheel

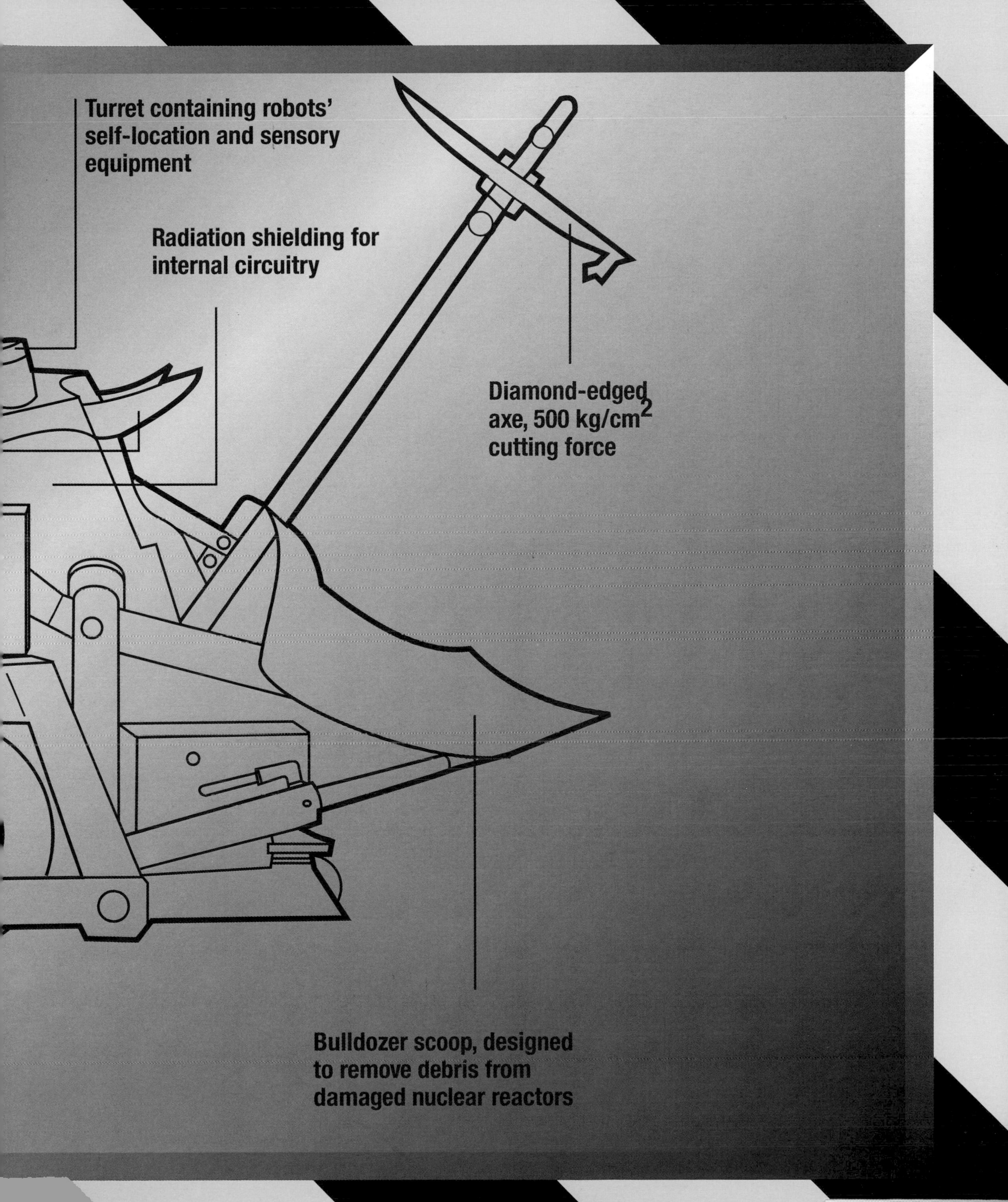
Turret containing robots' self-location and sensory equipment
Radiation shielding for internal circuitry
Diamond-edged axe, 500 kg/cm2 cutting force
Bulldozer scoop, designed to remove debris from damaged nuclear reactors

COUNTER-CLAIM BY THE BBC VISUAL EFFECTS DEPARTMENT

Shunt's incredibly aggressive profile reflects the VED's decision to create a robot that could be successfully deployed in the sumo-wrestling challenge. The resulting design is reminiscent of a steroid-enhanced bulldozer, with enormous pushing power and very low sides and ground clearance to ensure that nothing can slide underneath and up-end him – especially during the Sumo bouts in the Gauntlet. However, there was also a need for a more active weapon, and since the preferred chainsaw was rendered unfeasible by the mass of metal plating at front and rear, a diamond-edged axe was added to Shunt's rear section, just above the bucket scoop. The rear-mounted bucket scoop and the axe are powered by CO_2 cylinders.

Shunt's chassis and scoops are made of steel, while his superstructure is moulded fibreglass. He is powered by a 12-volt motor, and took five weeks to build. His extremely resilient steel structure and low profile ensure that he can take just about anything thrown at him by the competitors' robots. However, Shunt's designers were, initially, a little worried about the threat from the powerful axe wielded by the competitor robot Killertron, but Shunt acquitted himself admirably against this threat, and remains one of the most serious problems facing competitors in *Robot Wars*.

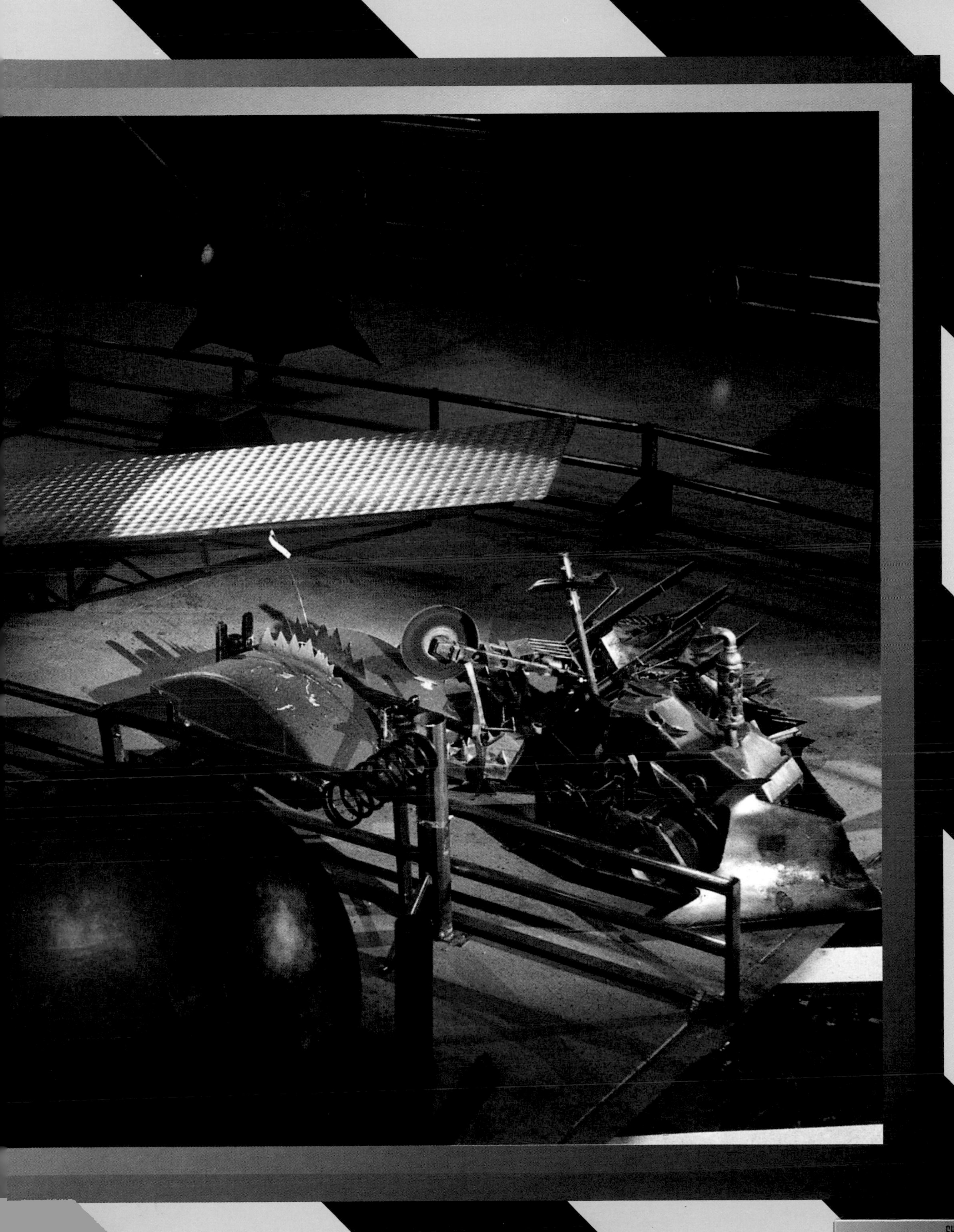

MATILDA

Matilda, who looks like a hybrid cross between a tank and a dinosaur, is a particularly fearsome and cantankerous machine. Solidly built, low and squat, she has a segmented back, from the rear end of which protrudes a vicious, moveable chainsaw. Her horned head, which includes a neck-frill reminiscent of a Stegosaurus, is dominated by a pair of powerful hydraulic tusks or mandibles, which she uses to pierce and mangle her opponents.

Matilda was originally discovered by a team of oceanographers in a deep underwater cave off the eastern seaboard of the United States. It was initially suggested by the team of scientists and archaeologists assigned to study her that Matilda might be the product of some blasphemous, long-forgotten science, a relic of an unknown but highly advanced prehistoric civilization, of which virtually no trace today remains. Subsequent research seems to have reinforced this theory. Certain inscriptions have been discovered within Matilda's internal circuitry, inscriptions which bear a passing resemblance to Sanskrit (until now considered to be the first written human language), but apparently pre-dating it by an unknown number of centuries. In a leap of inspired intuition, the leader of the research team (who, for security reasons, must remain nameless) suggested that Matilda might actually be a surviving artefact from the legendary continent of Atlantis.

This view was met with derision from the scientific community, until it was pointed out that, according to legend, Atlantis was destroyed in a violent cataclysm of some kind. Some commentators have maintained over the years that the Atlanteans were actually obliterated as a result of irresponsible misuse of their fabulous technology. Could the robot known today as Matilda actually be an example of that technology?

WEIGHT: 116 KG

LENGTH: 140 CM

WIDTH: 66 CM

HEIGHT: 66 CM

MAXIMUM SPEED: 13 KPH (8 MPH)

POWER: RECHARGEABLE BATTERY UNIT

WEAPONRY: HYDRAULIC TUSKS; 3000 RPM CHAINSAW-TAIL

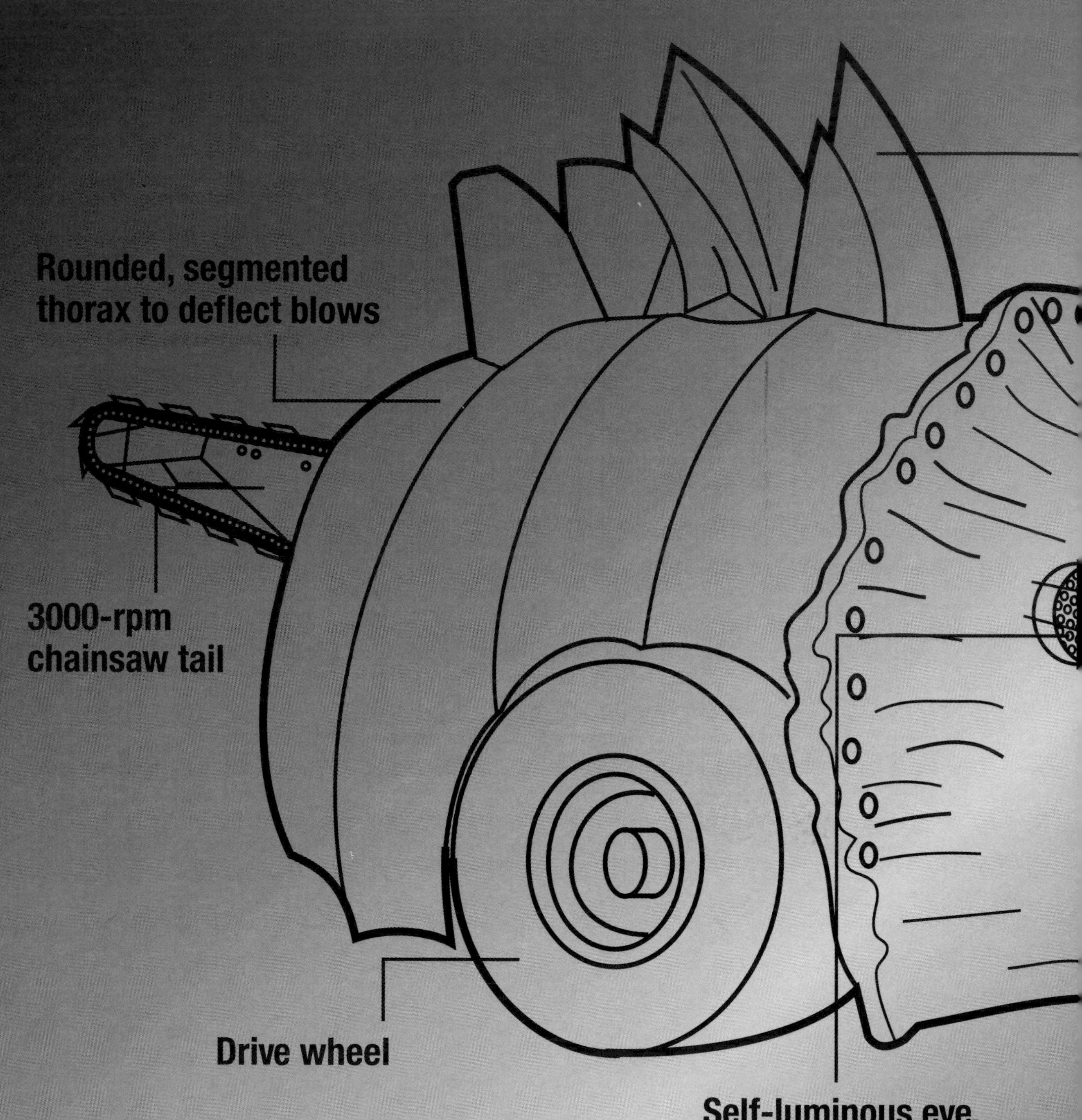
Rounded, segmented thorax to deflect blows
3000-rpm chainsaw tail
Drive wheel
Self-luminous eye, enabling robot to locate enemies in poor light

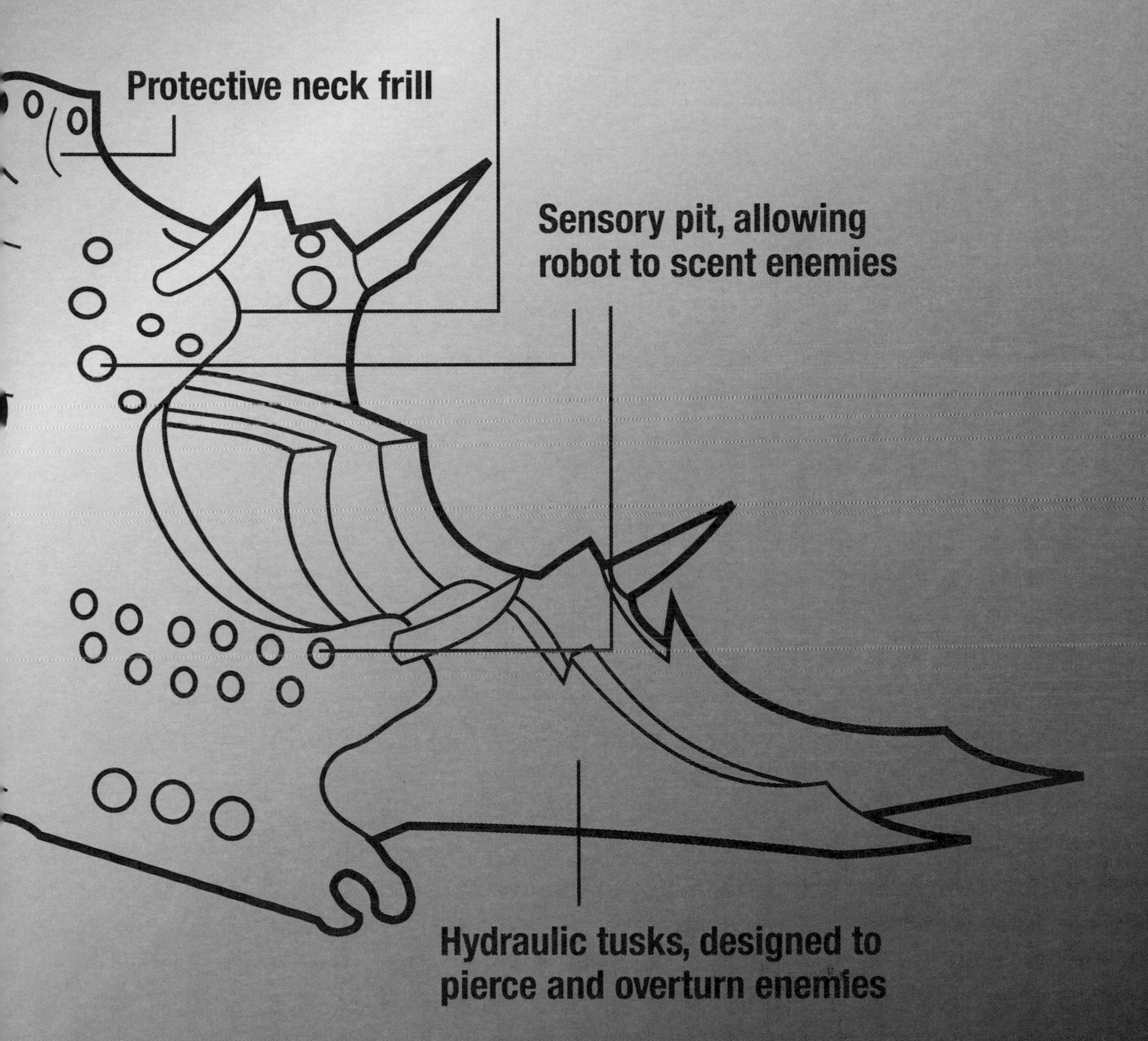
Protective armoured dorsal spikes
Sample of internal head circuitry carbon-dated to 10,000 years old
Protective neck frill
Sensory pit, allowing robot to scent enemies
Hydraulic tusks, designed to pierce and overturn enemies

Is it possible that the Atlantean civilization was destroyed by marauding robots such as Matilda?

Unlike her robotic partners in crime, Matilda has the appearance of a biomechanical creature, with an internal circuitry based on the hormones oestrogen and progesterone, which combine to regulate her aggression levels during combat. However, Matilda's performances during the First and Second Wars would suggest that those aggression levels can easily surge beyond all control.

It is unlikely that we will ever know for certain where (or when) Matilda came from, unless, of course, we can find a way of communicating with her. (All attempts at this have, so far, resulted only in serious injury to the investigating scientists.) One additional fact, however, remains: a tiny piece of the hormonal circuitry within the robot's brain has recently been carbon-dated, revealing her to be in excess of 10,000 years old ...

A set of extremely unpleasant hydraulic tusks at the front and a 3000-rpm chainsaw at the back make Matilda practically invulnerable, as was proven at least twice in the First Wars. The first occasion saw her about to dispatch Barry (a robot built by 17-year-olds Amy and Daniel from the Hagley Roman Catholic High School), her fiendish aggression was frustrated only by the lack of time before the end of the round. On the second occasion, Matilda neatly sliced the radio control mast from the back of Scarab during the grand final of the First Wars.

COUNTER-CLAIM BY THE BBC VISUAL EFFECTS DEPARTMENT

Designing and building Matilda was particularly challenging for the VED, especially since supervisor Chris Reynolds had the innovative idea of making her appear to be part machine and part alien dinosaur. Initial concept sketches proved to be too uninteresting and buglike, as did the first two models, which were constructed of clay, covering a polystyrene framework. It was very important to Chris and his team that Matilda should have a really definite personality that would set her aside from the other house robots: a different kind of menace ... the menace of the wild animal, as opposed to that of the implacable war machine.

In the nick of time, the team came up with the brilliant designs for the hideous entity now seen in *Robot Wars*, a ravenous, scaly, purple-silver monstrosity with eyes as expressionless as a Great White Shark's. She took five weeks to build.

Matilda's body is fashioned from fibreglass matting, which is more resistant to heavy blows than steel. She is driven by a 12-volt motor (with 80 amp speed controllers), which is to be upgraded to 24 volts to improve her speed. Her tusks are operated by a powerful pneumatic CO_2 system, while a 76 cc motor drives her steel chainsaw tail which, perhaps in keeping with Matilda's foul temper, has a habit of constantly throwing its chains.

Matilda's greatest strength is the extreme toughness of her skin, which is capable of resisting just about anything that is thrown at it. Her outer structure is smoother and less complicated than her robotic colleagues, which means that attacking blows from competitors' robots simply glance off. If she has any weakness, it is probably a lack of speed, although this is to be rectified with the upgrade in her motor.

DEAD METAL

Dead Metal is perhaps the most bizarre-looking and terrifying of the house robots, not least because fully half of his bulk is taken up by a pair of gigantic pincers, giving him the appearance of a hideous and deadly mechanical insect, a cross between a scorpion and an industrial workstation. Once these pincers have grabbed a victim, there is little hope for the robot, as a 3000-rpm circular saw mounted on an elongated boom swings up and over the top of the house robot, to crash down upon his hapless opponent.

Unlike the other house robots, Dead Metal possesses a steel exoskeleton composed of jagged armour plates with spiked struts covering his wheels. This extremely unusual design has resulted in some rather wild speculation as to his origins, with some commentators suggesting that he may not be from this planet. This is unlikely, however, since analysis by investigating scientists has revealed the steel in his body to be consistent with current steel production on Earth.

As with his robotic colleagues, it is something of a mystery just how Dead Metal came to be a participant in *Robot Wars*, but recently there have been a number of sinister rumours circulating on certain conspiracy-oriented Internet websites. (Don't try to look for these websites: all those dealing with the origin of the house robots have mysteriously vanished from the Internet.) According to one assertion, made by an anonymous person claiming to have worked for the United States Defence Advanced Research Projects Agency (DARPA), an experimental electrodynamic time machine, built in a top secret facility somewhere in the southwestern USA, proved disastrously successful. A doorway was opened on to the distant future, sucking Dead Metal, along with one of his victims, into our own time. Fortunately for the scientists operating the machine, the robot was immediately subdued by the intensely powerful negative-time field present in the doorway.

The victim, apparently human, was mortally wounded, but before he died, he managed to explain that he was a political dissident, and that the robot belonged to the Berserker Empire of 16,000 AD. The Berserkers are (or will be) cybernetically enhanced humans holding half the galaxy in their totalitarian grip. Thermonuclear-powered robots such as Dead Metal are the front-line invasion force sent to each

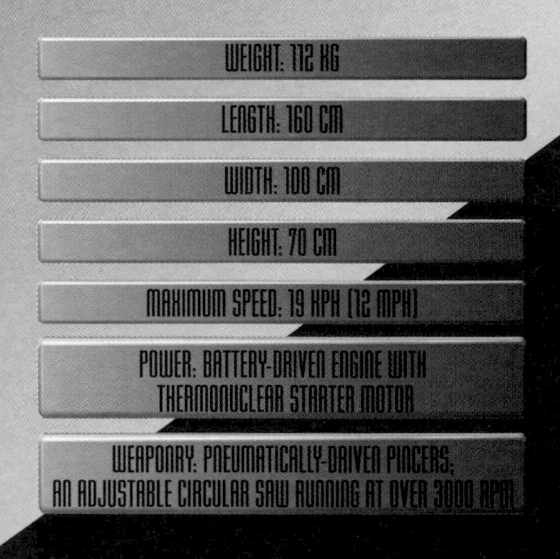

target planet, destroying all resistance in the most brutal manner imaginable.

If these claims are true, then it seems likely that Dead Metal is still the worse for wear after his impromptu trip through time, which can only be good news for the other robots pitted against him in *Robot Wars*, not to mention the rest of the human race – for the time being anyway!

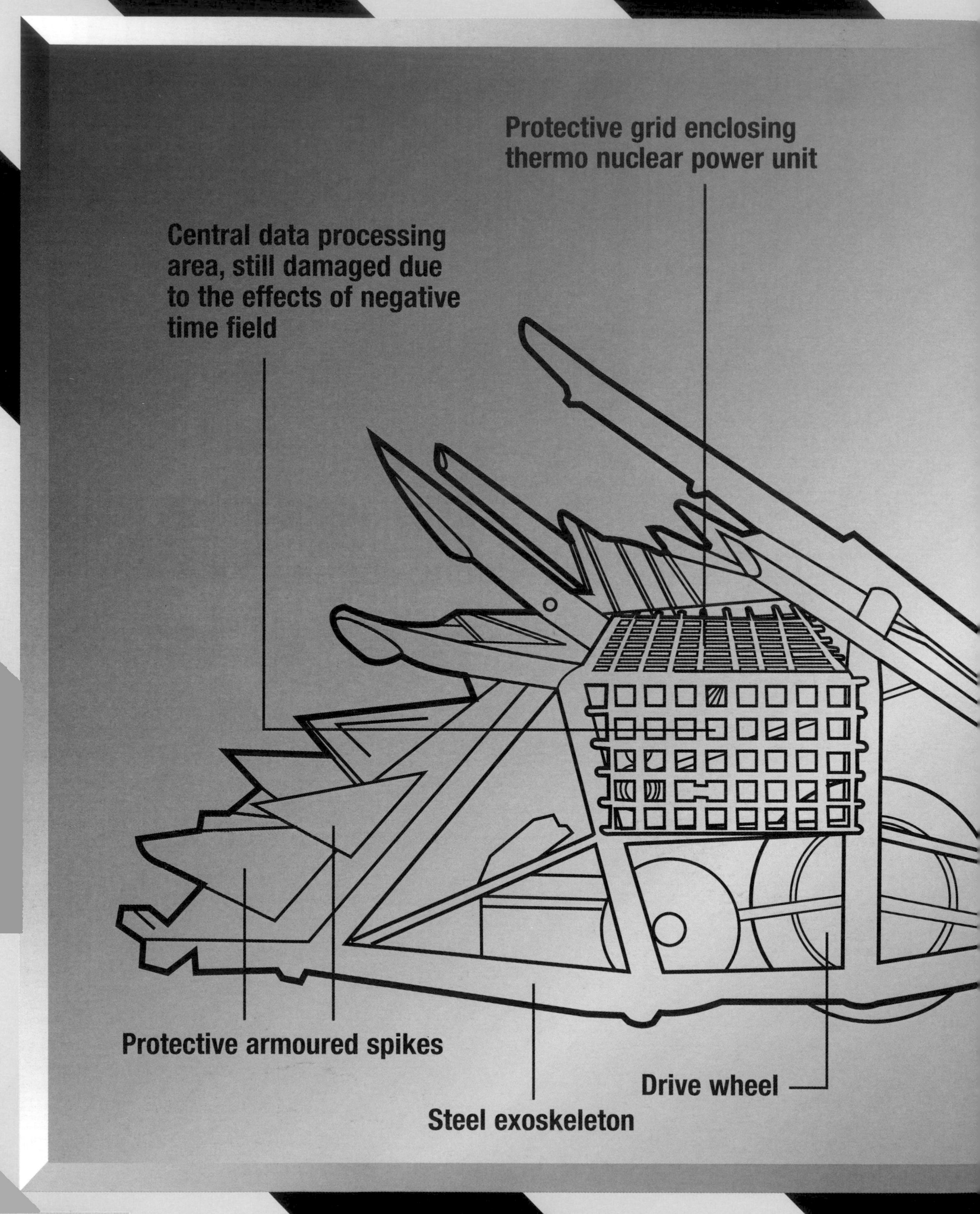
Protective grid enclosing
thermo nuclear power unit
Central data processing
area, still damaged due
to the effects of negative
time field
Protective armoured spikes
Drive wheel
Steel exoskeleton

COUNTER-CLAIM BY THE BBC VISUAL EFFECTS DEPARTMENT

Of all the house robots, Dead Metal has presented the VED team with the most problems. Right from the beginning, his jaws didn't work quite as well as had been hoped, and he had to be rebuilt with the increased power of a 24-volt motor. In addition, the motor driving his circular saw is not powerful enough, and is to be upgraded.

Dead Metal's distinctive exoskeleton has also presented problems: his internal wiring is not as well shielded as that of the other house robots, and has a habit of working loose during the more extreme battles. According to Chris Reynolds, the ambitious design of Dead Metal meant that the VED tried to do too much with him in the First Wars. The result was that he failed to live up to his full potential. Sadly his operating characteristics did not match the visual brilliance of his design.

For this reason, Dead Metal went through a period of redesign and technical refinement in preparation for the Second Wars, including changes to the specifications for his circular saw arm, which now moves through an arc of only 80° instead of 180°. The internal linkage system is also more powerful, and the old cutting disc has been replaced with a diamond disc. These improvements combine to vastly increase the power of Dead Metal's primary weapon. In addition, the power of Dead Metal's jaws has been increased, and his bodywork has been given a more rusty, grungy look. Not only is Dead Metal more powerful than he was in the First Wars, he's a whole lot uglier, too.

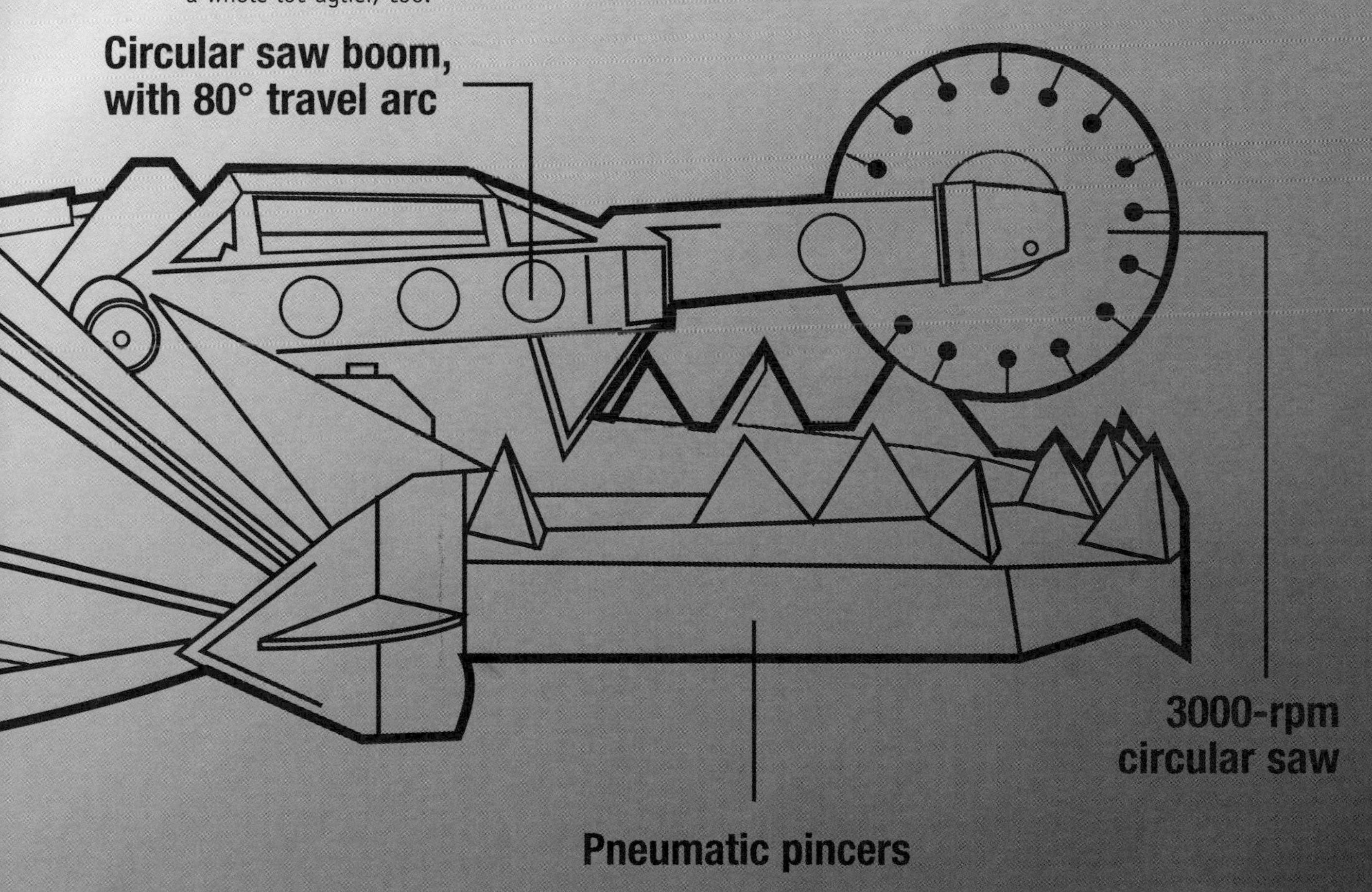

SERGEANT BASH

Sergeant Bash is a camouflaged, militaristic nightmare. His tank-like forward section is fronted by an angular, razor-sharp ram, while his rear section, with its flared wheel-cowling proudly displaying his sergeant's stripes, contains the lethal flame-thrower turret. He also carries a powerful circular saw, just for good measure.

Over the years, many outrageous rumours have circulated regarding the top-secret military test site known as Area 51 in Nevada. Most of them centre around the idea that the US Government has captured several alien spacecraft which allegedly have crashed in the area, and has 'reverse engineered' them to produce its own highly sophisticated aircraft and weapons systems. While these rumours may or may not be true, the fact remains that some very powerful technology is being developed at Area 51, including the F-117A Nighthawk fighter plane and the Northrop B2 Stealth Bomber.

In the late 1960s, ranchers in southwestern USA began to suffer the attentions of a mysterious predator that killed and mutilated their livestock, horses, and domestic animals. These so-called 'animal mutilations' have perplexed investigators of the unexplained ever since. No perpetrators have ever been apprehended, although the mutilations numbered in the hundreds every year. In each case, certain organs were found to have been removed, and the carcasses displayed strange burn-marks. Many ranchers reported seeing black, unmarked helicopters in the area just before and after a mutilation, leading some to speculate that the military might have been conducting some kind of ultra-secret weapons experiment on the hapless farm animals. Other people in the area claimed to have seen a small, unusual-looking object speeding across their fields at ground level. When asked by investigators to draw what they had seen, they invariably came up with a similar-shaped object – a squat, lozenge shape with two circular features towards the rear, which may or may not have been wheels.

This only added to the mystification of researchers investigating the mutilations, especially the ufologists who had assumed that they had been carried out by sinister aliens conducting some kind of genetic engineering experiment. It was not until 1990 that someone managed to catch the mysterious mutilator on video, revealing it to be a wheeled vehicle, apparently equipped with a powerful heat weapon mounted on its dorsal surface. Unfortunately, very few people have seen this astonishing video, since the unnamed journalist who shot the sinister footage immediately received a visit from three men claiming to be military

WEIGHT: 120 KG

LENGTH: 140 CM

WIDTH: 90 CM

HEIGHT: 90 CM

MAXIMUM SPEED: 13 KPH (8 MPH)

POWER: FOUR BATTERIES RUNNING IN PARALLEL. CONDUITS ATTACHED TO FLAME-THROWER VENT DIRECT HEAT TO POWER STEAM ENGINE

WEAPONRY: PROPANE-FUELLED FLAME-THROWER MOUNTED ON A 360° TURRET; STEAM DRIVEN CIRCULAR SAW RUNNING AT 3200 RPM

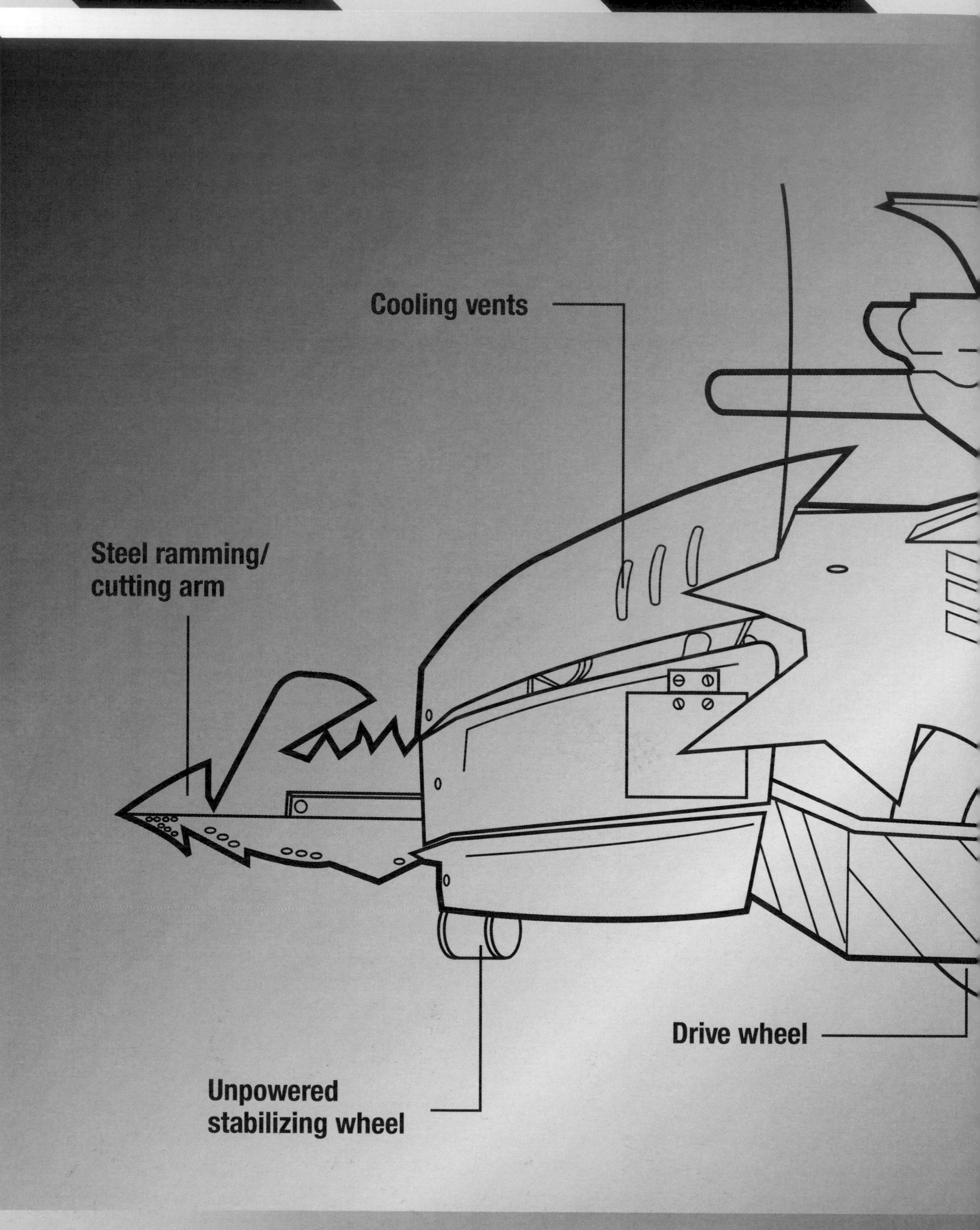
Cooling vents
Steel ramming/
cutting arm
Drive wheel
Unpowered
stabilizing wheel

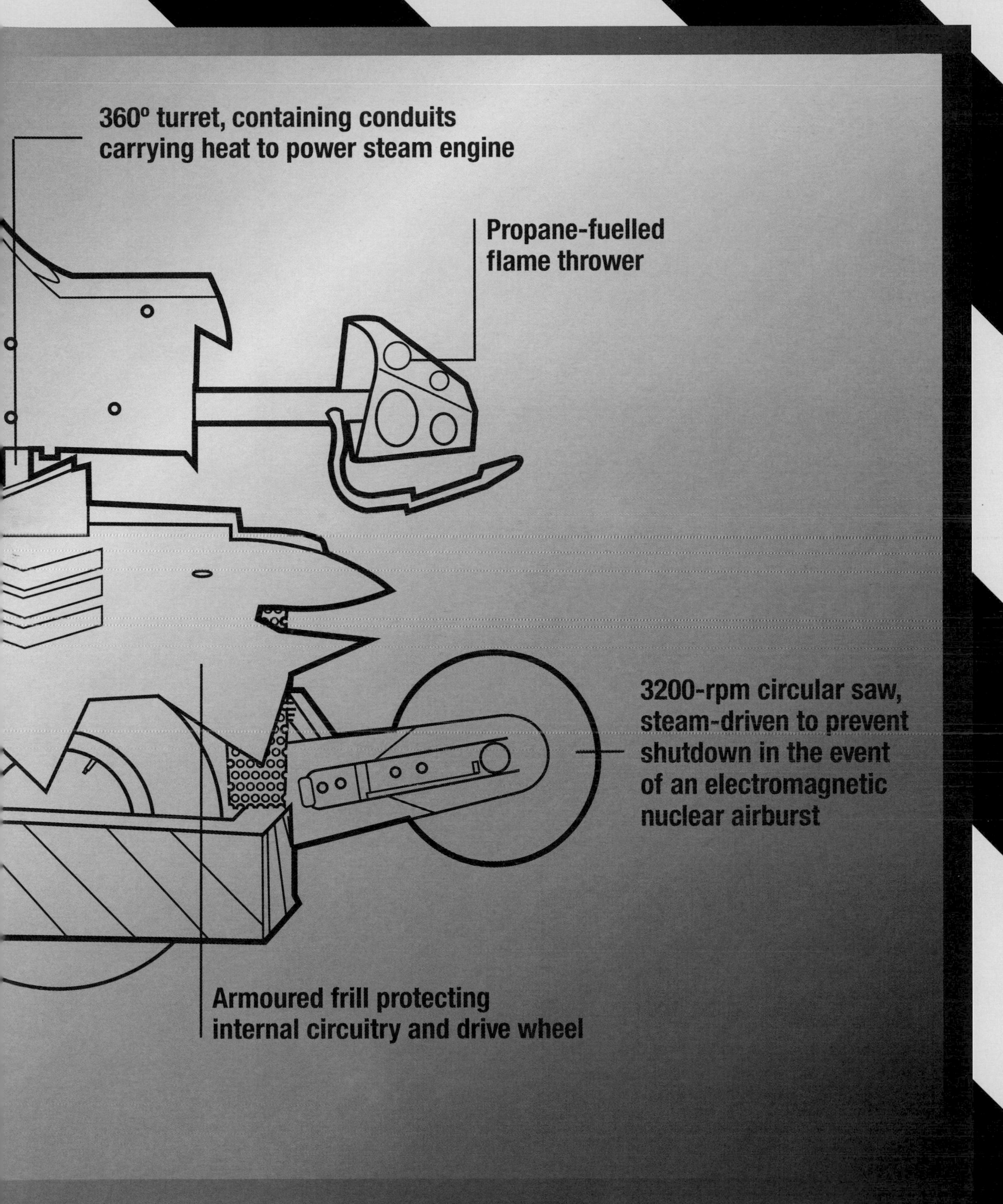
360º turret, containing conduits carrying heat to power steam engine
Propane-fuelled flame thrower
3200-rpm circular saw, steam-driven to prevent shutdown in the event of an electromagnetic nuclear airburst
Armoured frill protecting internal circuitry and drive wheel

intelligence agents. These men confiscated the video tape and warned the journalist to drop his investigations into the mutilation phenomenon. The journalist subsequently left his job and joined a New-Age commune just outside San Francisco. These days he is often seen wandering the streets of the city, shouting that the end of the world will soon be upon us.

As to the nature of the bizarre vehicle in the video, it was subsequently confirmed by an anonymous physicist who had worked for some years at Area 51 that the machine had been designed and built there. It seems that a top secret contract had been issued to the military by the Federal Emergency Management Agency (FEMA), to build an urban pacification device, code-named Sergeant Bash, for use in times of national emergency.

There have been a number of rumours regarding FEMA circulating among conspiracy theorists, the most common of which is that the agency is, in effect, a 'secret government' planning to declare martial law in the United States and turn the country into a totalitarian state. It has been suggested that one of the ways in which this may be achieved is through the use of the Sergeant Bash pacification droid, which would patrol each city, instantly eliminating all resistance.

While this can (possibly) be dismissed as one of the many paranoid – and often downright insane – scare stories floating around, it can be said with a fair degree of certainty that one or more Sergeant Bash droids have been active in the southwestern United States over several years, probably being released and monitored by military personnel in the frequently sighted unmarked helicopters.

However, the question remains: how did a Sergeant Bash droid wind up participating in *Robot Wars*? The likeliest answer would be that its designers have decided to conduct a disinformation campaign with the intention of drawing the public's attention away from Sergeant Bash's true purpose and origin in Area 51. After all, as the saying goes, the best place to hide something is in full view ...

COUNTER-CLAIM BY THE BBC VISUAL EFFECTS DEPARTMENT

It was important to the Mentorn that the VED should devise a way of constructing a flame-thrower that could add to the drama of the Wars, and yet would be controllable and safe. The producers were absolutely right, and Sergeant Bash has become one of the most feared opponents in *Robot Wars*. Also essential was a rear-mounted weapon that would deter the more courageous competitors from mounting an assault from behind, and for this reason the 3200-rpm circular saw was added.

The Sergeant's weapons design and colour scheme reflect his military aspirations; he is also distinguished from the other house robots by his large Rotavator drive wheels (a Rotavator being a machine for digging up gardens). In fact, Sergeant Bash contains more complex equipment than any other robot: his sophisticated weapons systems require the inclusion of gas bottles to power the flame-thrower, in addition to valves and CO_2 bottles. The components powering the flame-thrower and circular saw have increased the Sergeant's weight to a whopping 120 kilograms. However, his tremendous bulk has cost him in terms of power, and his 12-volt motor is to be upgraded to 24 volts in order to rectify this weakness.

Sergeant Bash is constructed of steel plating with an aluminium chassis, and took five and a half weeks to build. According to Chris Reynolds, one of the major problems encountered by the Sergeant during the Wars was interference from other radio control units during battle, which had the unfortunate (and potentially fatal) effect of shutting down the flame-thrower. The house roboteers solved this problem, however, by insulating the Sergeant's radio control receivers with a wrapping of silver foil.

Sergeant Bash is also potentially weakened by his extremely long wheelbase (another side-effect of the huge mass of internal components). This has the effect of compromising his balance and stability, and leaves him vulnerable to overturning by some of the more powerful competitors' robots. For this reason, the Sergeant is scheduled for a major overhaul some time in 1999. When that happens, Sergeant Bash will become one of the most powerful and dangerous participants in *Robot Wars*.

SIR KILLALOT

According to information retrieved from the Internet, after the terrifying success of their first experiment in time travel – when they recovered Dead Metal – DARPA began to alter the design parameters of their electro-dynamic time machine, in accordance with the Many Worlds Interpretation of the Danish physicist Niels Bohr. Their intention was to gain access to one or more of the infinite number of parallel universes said to co-exist with our own within the same spatial co-ordinates.

Once again, the ludicrously irresponsible scientists at DARPA were successful, opening a gateway to one of the ghost universes. And once again, their success exposed them (and the rest of our world) to extreme danger. Even their experience with the machine now known as Dead Metal failed to prepare them for what emerged from the temporal gateway.

At first, they assumed it to be some kind of tank or other armoured vehicle, since it moved on caterpillar tracks, and its form was dominated by two horrendous weapons, a lance-like flame-thrower and a vicious cutting arm. As the monstrosity finally exited from the spinning toroids of the time machine and came to a halt in the laboratory, the technicians realized that it also had a helmeted head, from which shone two baleful, red eyes. According to one of the technicians – an informer who made this information available before disappearing in mysterious circumstances – everyone present had the distinct impression that this hideous mechanical creature, regarding them impassively from the centre of the laboratory, was trying to decide what to destroy first.

However, like Dead Metal, this new arrival seemed to have suffered a serious loss of power by virtue of its unexpected and traumatic journey between universes. Realizing what had happened, the technicians took the opportunity of attempting to establish

WEIGHT: 200 KG

LENGTH: 120 CM

WIDTH: 120 CM

HEIGHT: 130 CM

MAXIMUM SPEED: 8 KPH (5 MPH)

POWER: 2 X 1-HORSEPOWER 24-VOLT MOTORS

WEAPONRY: FLAME-THROWER LANCE;
PINCERS – POWERED BY AN 8-HORSEPOWER HYDRAULIC PUMP

communications with the robot. This they succeeded in doing, and discovered to their astonishment that its universe, the legendary continent of Atlantis was not destroyed in 10,000 BC, but continued its domination of the world. As a consequence, the First World War did not begin in 1914, but in 1150 AD, around the time of the Crusades in our universe.

This robot, which called itself Sir Killalot, claimed to be an autonomous fighting unit for use on the battlefields of medieval Europe, which was the theatre of war between Atlantis and its great adversary, Lemuria (another legendary civilization, thought to have fallen thousands of years ago in our universe).

Recruited through unknown means to do battle in *Robot Wars*, Sir Killalot just about tolerates the other house robots, although he seems to have some affinity with Matilda, which would support the theory that she herself originated in Atlantis 10,000 years ago.

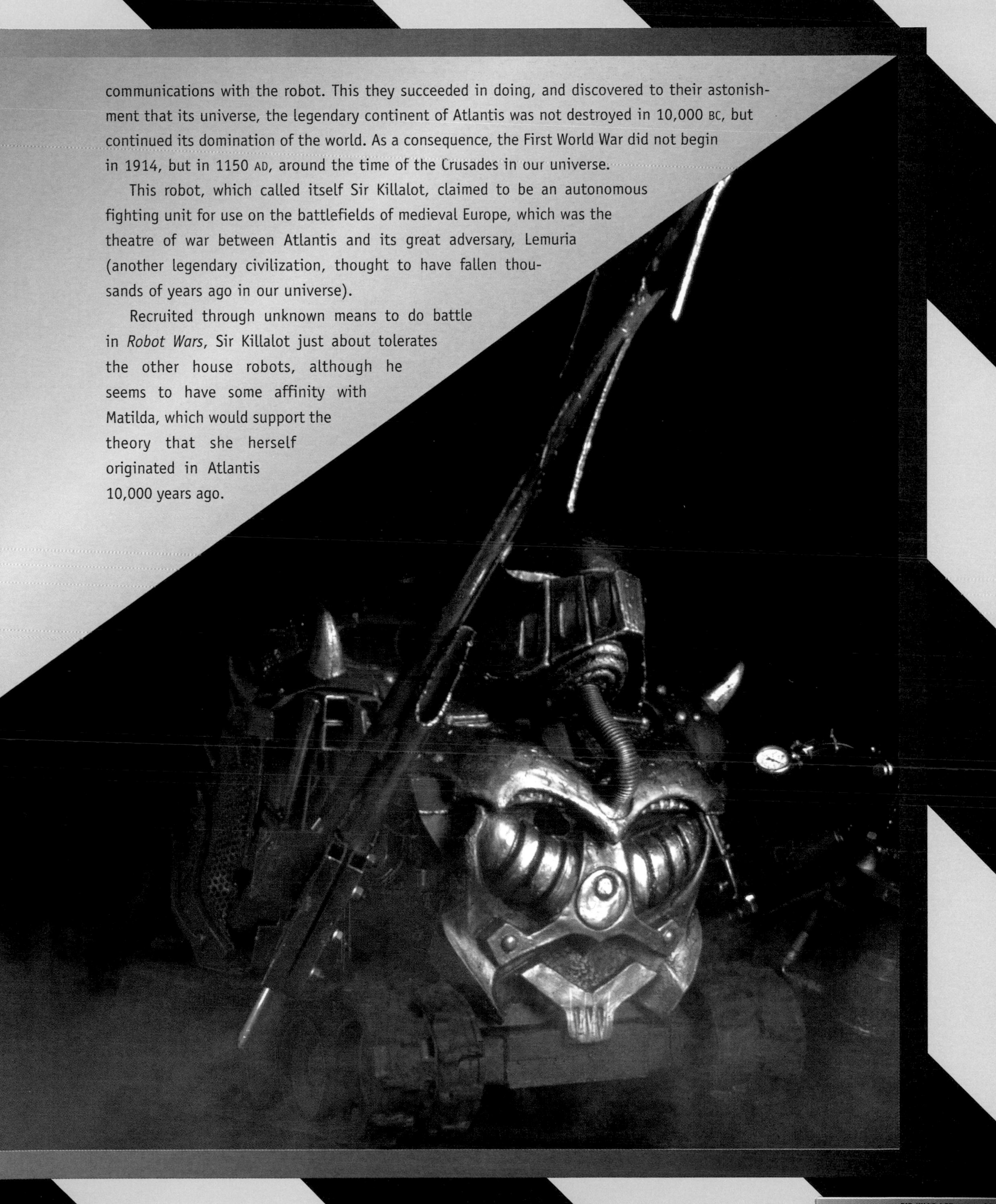

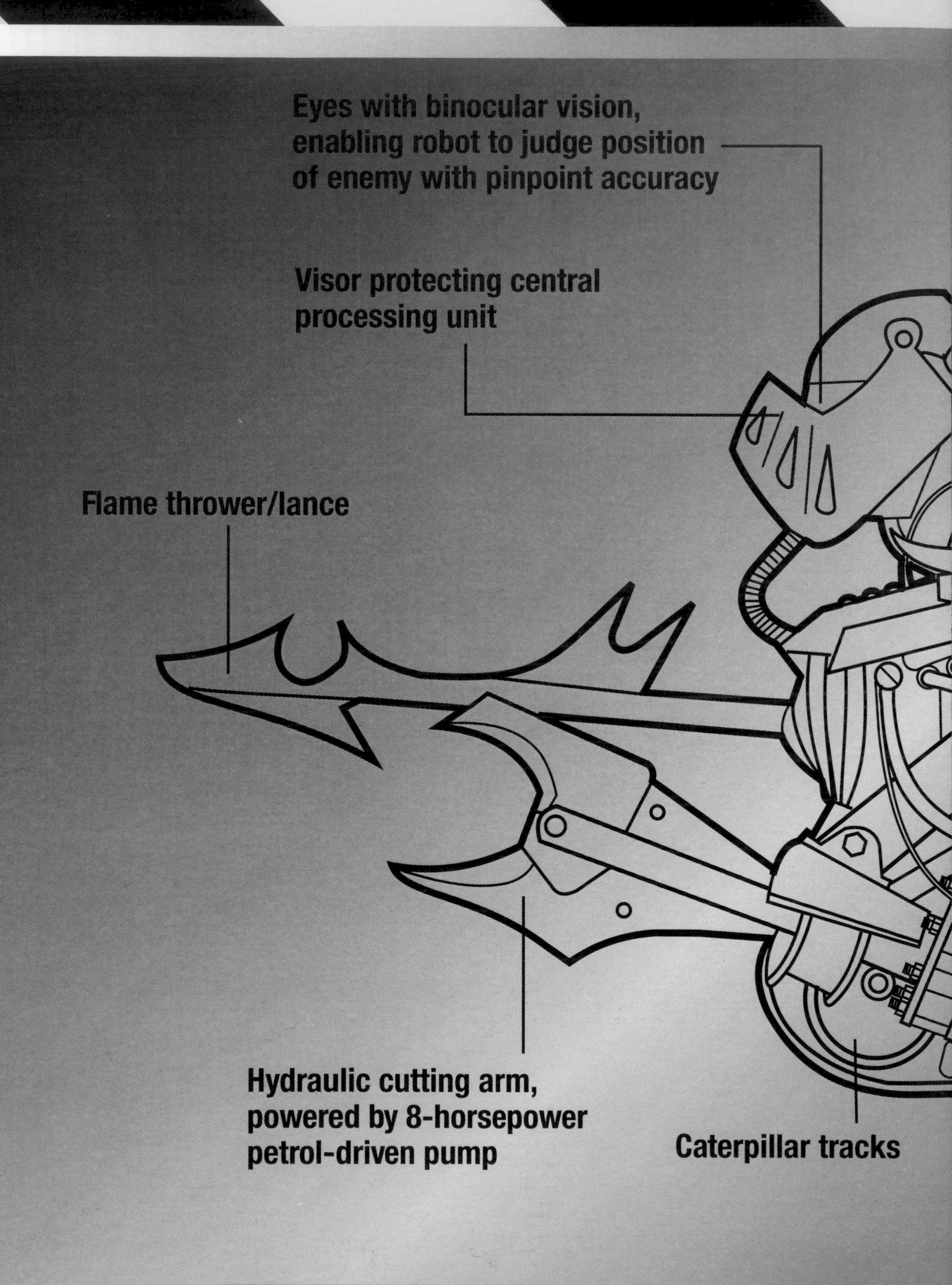
Eyes with binocular vision, enabling robot to judge position of enemy with pinpoint accuracy
Visor protecting central processing unit
Flame thrower/lance
Hydraulic cutting arm, powered by 8-horsepower petrol-driven pump
Caterpillar tracks

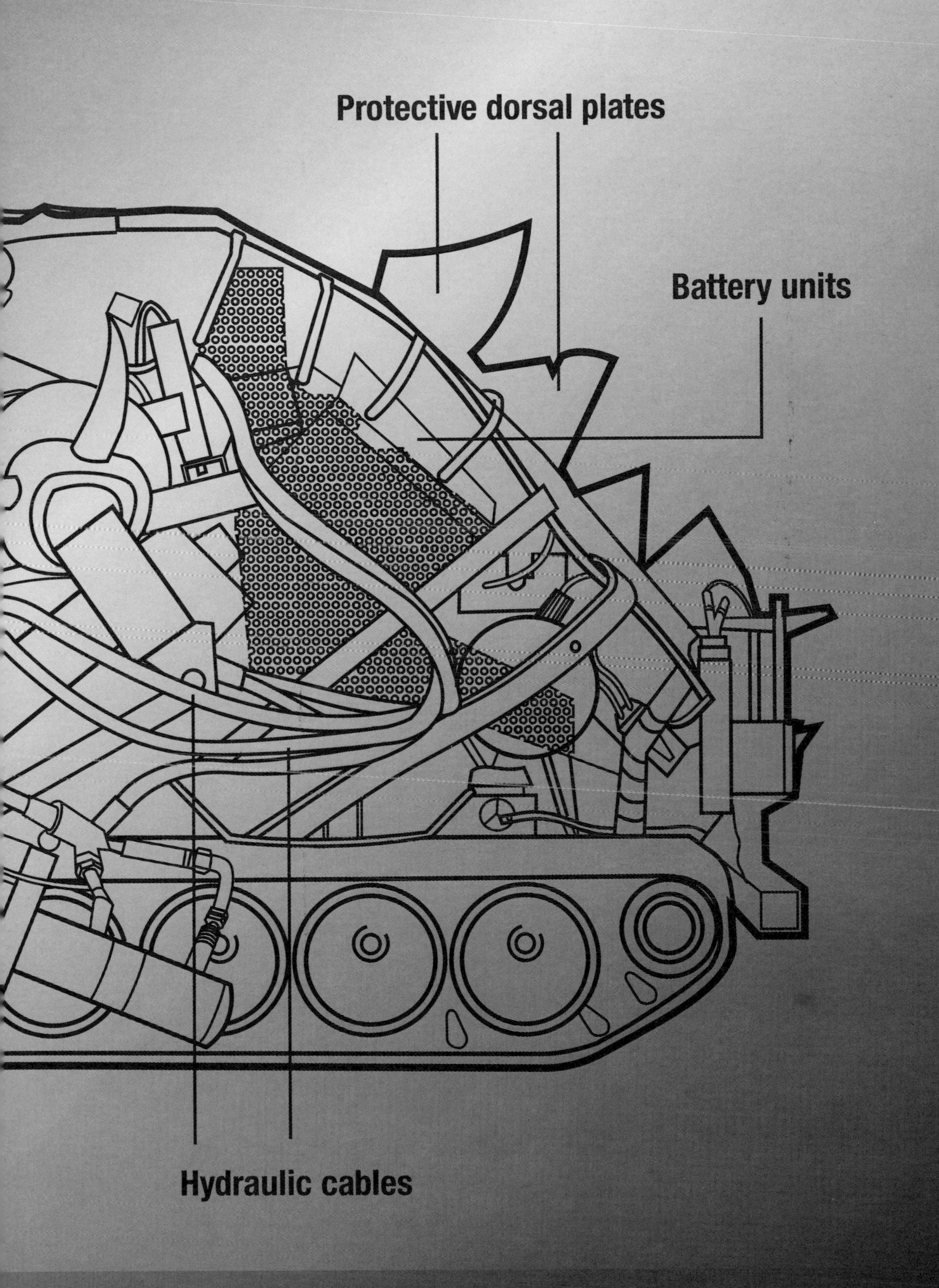
Protective dorsal plates
Battery units
Hydraulic cables

COUNTER-CLAIM BY THE BBC VISUAL EFFECTS DEPARTMENT

Sir Killalot is the latest – and by far the most impressive – addition to the violent pack of house robots. Considering the complexity of his design and the formidable nature of his weaponry, Sir Killalot was built in an astonishingly short time – about three weeks. Equally surprising is the fact that his design and construction proceeded with no major problems, apart from his speed controllers. For this the construction team used a spare set of controllers from the other house robots. However, installation in Sir Killalot pushed the controllers to the edge of their capacity, and they developed a habit of overheating. Once these were replaced with new controllers, and a large cooling fan was added, the problems with overheating disappeared.

According to Colin Tilley of the VED, one of the design elements that make Sir Killalot such an extreme danger to the competitors' robots is the power and sophistication of his hydraulics. His main weapon is a hydraulic cutting arm, of the type used by rescue services to cut people out of the wreckage of crashed cars and so on. However, the VED team decided to go one better with the cutting arm, which they dismantled and uprated, replacing the original hydraulic ram, which exerted 8 tonnes of cutting force, with one that exerts 15 tonnes of force. The increased power of the hydraulic ram also means that Sir Killalot's weapon cuts much faster than the version used by the rescue services, who need a slow-moving tool that they can reverse quickly when cutting close to a human being. Sir Killalot's cutting arm opens and closes much faster, and is easily capable of slicing through the armour plating and even the axles of any robot unlucky enough to find itself the object of its dastardly attentions.

According to Commander-in-Chief Craig Charles, Sir Killalot has an added advantage in his height (at 130 centimetres, he is 40 centimetres higher than Sergeant Bash – the tallest of the original house robots). As a consequence of this Sir Killalot can obscure a roboteer's line-of-sight, making it extremely difficult to control his or her machine. One of the reasons for this is that, when your robot is facing you, the control movements are reversed: to get your machine to move towards the left, you must push your steering control to the right, and vice versa. In the heat of battle, this can be confusing enough in itself, without having your robot completely obscured by a gigantic lump of metal that is intent on ripping it to shreds.

For those prospective roboteers who are thinking that Sir Killalot will put a swift end to their efforts, Colin Tilley has some words of encouragement ... not a great deal of encouragement, but still worth considering. While the speed with which Sir Killalot was built is a tribute to the technical abilities of the VED, this does nevertheless leave him with one potentially dangerous weakness. According to Colin, the robot's internal components are quite exposed, and it would be quite easy to knock out a hydraulic line, a battery (located on his back) or a battery lead. However, he adds, this is easier said than done!

BUILDING YOUR OWN ROBOT

Any machine capable of standing up to the might of the house robots, not to mention the many other obstacles to be encountered during *Robot Wars*, will have to combine successfully a number of different elements. The purpose of this chapter is to guide the reader through the requirements of robot design, using a number of competitors' robots as examples. The rules and guidelines governing *Robot Wars* in Europe follow, as far as possible, those in America, where the sport originated. It is important to maintain internationally recognized regulations, since future wars may well include robots from across the globe. Before attempting to build a robot it is essential that you join the Robot Wars Club in order to obtain the official rules and safety regulations (see page 141).

One of the best aspects of *Robot Wars* is that it encourages teamwork and co-operation between individuals (although single-member teams are also more than welcome): for this reason we have seen robots designed by families, friends, schools, technical colleges, and workplaces. Each team usually includes one person who is responsible for the basic design of the robot, one who specializes in electronics, and another who takes care of the mechanical side of things. Of these, at least one should be a competent robot driver. Although any number of people can be involved with the design and construction of the robot, there is an upper limit of three crew members for actual participation in the show: the main reason for this is that the pit area has limited space. However, other members of the design and construction team are given prime spots in the audience.

TYPES OF ROBOT

Although all the robots seen in *Robot Wars* are basically radio-controlled fighting machines, closer inspection will reveal a wide variety of different types, based on the particular needs envisaged by the designers and builders. Before you even put pen to paper for the first design sketches, you need to consider exactly what you want from your machine. Among the first things to consider are the weight, the type of drive mechanism, bodywork (which can play a crucial role in defence), manoeuvrability, and control mechanisms. It is important to remember that simply packing every single idea into one machine is unlikely to prove a successful strategy for two main reasons. Firstly, it will almost certainly push your robot over the maximum weight allowed; and secondly, the well-known law that says anything that can go wrong *will* go wrong applies to your robot, and the more features it contains, the more potential problems you will have when building it.

Compromise is therefore an essential element in a successful robot design. For example, low ground clearance may protect you from other robots intent on overturning you, but it may prove a serious problem when you try to negotiate a ramp; or, you may consider strength and power to be the most desirable attributes for your robot, but they may prove to be your undoing when you try to make a quick escape from a tricky situation, and find that you simply don't have the necessary speed.

However, it must be remembered that *Robot Wars* is a television show enjoyed by millions, and consequently the keyword is *spectacle*. For this reason, robots with elaborate weapons or other interesting design features are to be encouraged. It is always a good idea to include

Behemoth demonstrates the advantages of a lifting arm in a nail-biting encounter with Inquisitor.

COMPETITOR ROBOT

MEGA HURTS

HEAVYWEIGHT

CAPTAIN: DAVE LUND

ROBOTEERS: TONY LUND, GORDON KEELING

LENGTH: 115 CM

WIDTH: 60 CM

HEIGHT: 63 CM

WEIGHT: 71.8 KG

MAXIMUM SPEED: 6 KPH (4 MPH)

GROUND CLEARANCE: 40 MM

TURNING CIRCLE: ZERO

POWER: 2 X 12-V BATTERIES; 4 X FORD FIESTA WINDSCREEN WIPER MOTORS FOR LOCOMOTION, 3 WINDSCREEN WIPER MOTORS FOR WEAPONS SYSTEM

WEAPONRY: REAR-MOUNTED HACKSAW, WITH CHISEL

Cassius attempts to overturn Wizard, while Sir Killalot looks on approvingly.

The double-wedged Demolition Demon plays with fire.

something to make your robot stand out from the rest. For example, a rotating weapon. Chainsaws and circular cutting discs provide a lot of visual excitement, as do hammers pummelling the bodywork of an opponent. A particularly useful weapon (and one which is guaranteed to raise delighted cheers from the audience) is a lifting arm, whose purpose is to overturn your opponents and thus knock them out of the competition. A word of caution, however. More and more contestants are toying with the idea of including self-righting mechanisms in their designs, which can flip an up-ended robot back on to its wheels! You have been warned.

Another option is to avoid offensive weapons altogether, and to rely on the shape of your robot to guide it to victory. Such machines are usually known as 'wedges', due to their characteristic profile. The champion of the First Wars in Britain, Roadblock, was designed along these lines, as was the former American

COMPETITOR ROBOT

WHEELOSAURUS

HEAVYWEIGHT

CAPTAIN: PETER GIBSON

LENGTH: 130 CM

WIDTH: 100 CM

HEIGHT: 65 CM

WEIGHT: 67.5 KG

MAXIMUM SPEED: 16 KPH (10 MPH)

GROUND CLEARANCE: 125 MM

TURNING CIRCLE: ZERO

POWER: 2 X 12-V BATTERIES, RUNNING ON 24-V SYSTEM; 2 X WHEELCHAIR MOTORS.

WEAPONRY: ARMOUR PLATE APPENDAGES ON WHEELS, FRONT-MOUNTED POLE

Groundhog (left) and Loco size each other up.

champion La Machine. The main advantages of the wedge shape are that it can be very useful in overturning other robots, as well as allowing space for sophisticated drive mechanisms within the weight restrictions. Nevertheless, such designs do not make for particularly exciting viewing, and prospective roboteers are advised to give some thought to more interesting designs. This is why there are additional awards in *Robot Wars* for design, sportsmanship, innovation, engineering, and originality.

For the more adventurous roboteers, there are two additional design options guaranteed to turn heads – assuming, of course, that they can be made to work properly. Flying robots, like hovercraft, and those which are carried aloft by balloons (these are especially challenging from the point of view of control) are particularly impressive. Unfortunately airfoil designs (helicopters, for example) are not allowed for safety reasons. Another alternative is the walking robot, for which there is an increased weight allowance of 136 kilograms. This design may well be hampered by fragility, and may not be able to carry much in the way of weaponry, but will definitely impress the Judges when they come to consider the most original designs, and the best-engineered robots. However, whatever the problems of the legged machine may be, for many this remains the ultimate challenge in robot building.

But whatever type they are, all robots are composed of two basic elements: the chassis and the body. The chassis contains the drive, power supply, steering and radio-control equipment. Given the number of elements included, the chassis will undoubtedly be the heaviest part of your machine, and so will be the main stabilizing force. Stability is extremely important, since one of the most popular attacking strategies is to try to overturn opponents.

In designing his robot Wheelosaurus, Peter Gibson wanted to move away from the wedge and box shapes beloved by so many roboteers. Despite their success, he felt that there were far too many of them. Using this strategy he ended up with one of the most original designs seen in the Wars. His primary concern was that

COMPETITOR ROBOT

MILLY-ANN-BUG

HEAVYWEIGHT

CAPTAIN: GEOFF WARREN

ROBOTEERS: MARTIN DAWSON, BEN WEAVER

LENGTH: 140 CM

WIDTH: 62 CM

HEIGHT: 40 CM

WEIGHT: 53.0 KG

MAXIMUM SPEED: 8 KPH (5 MPH)

GROUND CLEARANCE: 55 MM

TURNING CIRCLE: 70 CM

POWER: 2 X 12-V BATTERIES TO POWER THE ELECTRONICS AND WEAPONS; 4 X LARGER CAPACITY 12-V BATTERIES, ONE FOR EACH MOTOR. MOTORS DERIVED FROM A CORDLESS DRILL

WEAPONRY: REAR-MOUNTED RETRACTABLE WEIGHTED SPIKE, FRONT-MOUNTED CIRCULAR SAW; WOODEN MANDIBLES

his robot wouldn't fall over, which renders the majority of machines utterly helpless. Peter decided to opt for very large wheels, mounted on either side of a basic mild steel pole chassis. This gave the main body of his robot an almost spherical profile, dominated by the two large wheels, which are actually salvaged from an old Victorian pushchair.

However, his initial design idea proved impractical, because the wheels were too close together, reducing stability and making steering very difficult. Peter solved

A member of the Milly-Ann-Bug team works on the robot's internal components.

this problem by installing a gyroscope from a model helicopter. He wanted it to be as simple as possible, for the logical reason that the simpler a design, the less there is to go wrong. This strategy is worth keeping in mind, particularly if, as for many roboteers, money is an important factor in building your machine.

Simplicity was also an important factor for brothers Rupert and Chris Weeks when designing their robot Tantrum, mainly because, according to Rupert, unusual ideas would have taken too long to realize. Having said that, they at first went for an ingenious and ambitious drive system including caterpillar tracks and a freely spinning triple-wheel assembly mounted on the front of the robot, to allow it to climb over obstacles. The tracks, made from pieces of car tyre, caused the robot to be far too difficult to control, so they were removed in favour of two large rear drive wheels (powered by two 50-watt wheelchair motors), although the front-mounted triple-wheel assembly was retained. The entire robot had been designed around the track system, so when the tracks were abandoned, the strength of the front wheel assembly had to be increased by using heavy duty castors. In addition, the team found that the speed controller had a habit of overheating, so they installed fans to keep it cool.

Geoff Warren and his team came up with a rather novel approach to robot design. They distributed a questionnaire soliciting ideas from their work colleagues. They then collated the results at a meeting, and picked the most interesting one. Initially, the complex design for Milly-Ann-Bug included three articulated sections, the intention being to have a machine reminiscent of a millipede. This design had good ground clearance and was difficult to topple over. However, its turning circle was considered too large, and the overall machine would have been too expensive and complicated to construct, so the team compromised and went for two sections instead of three.

Robo Doc has to be one of the most impressive looking machines seen so far in *Robot Wars*, in spite of (or perhaps because of) the trial-and-error way in which it was designed. According to team captain Mike Franklin, the main consideration was that Robo Doc should be something 'big and mean', not just another 'box on wheels'. So it was logical that wheels should be the first element to be abandoned in favour of tracks (which the team found in a scrap yard). The variable gearing in the robot's drive mechanism is set for 13 kph (8 mph), and its pulling power is over 360 kilograms. Robo Doc's main strength is that it can run either way up (so being overturned is no problem – unless, of course, it lands on its side). But it is also extremely robust, and can climb over obstacles up to 50 centimetres in height. If it has a weakness, it is that smaller robots can out-manoeuvre it. But Mike Franklin's ambitions in robot design do not stop with Robo Doc: in fact, he would like to go for the ultimate *Robot Wars* challenge, and build a legged machine.

The remote control units for one of the house robots.

COMPETITOR ROBOT

ROBO DOC

HEAVYWEIGHT

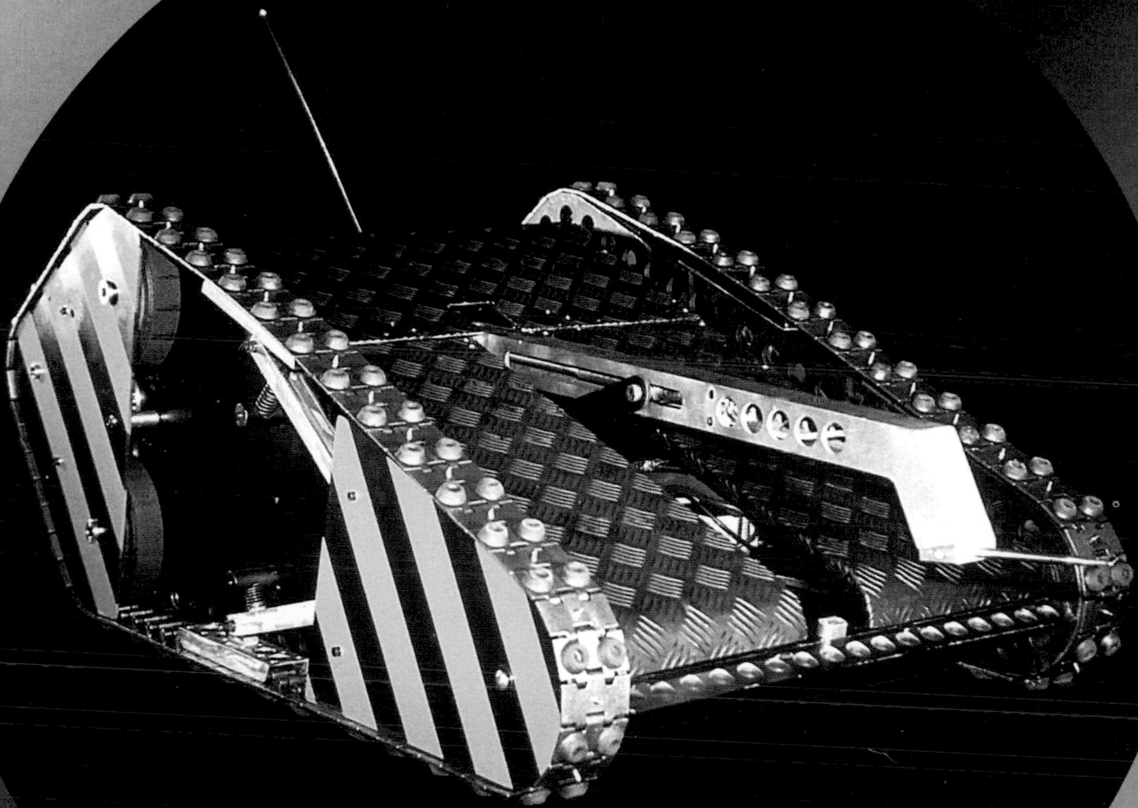

CAPTAIN: MIKE FRANKLIN

ROBOTEERS: JASON SMITH, PAUL JOHNSON

LENGTH: 160 CM

WIDTH: 78 CM

HEIGHT: 54 CM

WEIGHT: 79.9 KG

MAXIMUM SPEED: 24 KPH (15 MPH)

GROUND CLEARANCE: 50 MM

TURNING CIRCLE: ZERO

POWER: 2 X 12-V SEALED LEAD ACID GEL BATTERIES;
2 X 24-V 750-WATT MOTORS, DELIVERING 1 HORSEPOWER

WEAPONRY: FRONT-MOUNTED LIFTING ARM;
FRONT-MOUNTED CLEAVING KNIFE

THE DRIVE

The exact location of the drive unit within the chassis can make a lot of difference to your robot's performance characteristics. For example, a centre-mounted drive can reduce the turning circle to zero – meaning that your robot will be able to spin around while standing on the same spot. This high manoeuvrability, however, can result in decreased stability, and can make your machine vulnerable to overturning. It is well worth considering a more traditional drive position and chassis layout, with one wheel at each corner, like a car, which makes overturning much more difficult, but in turn will reduce manoeuvrability.

ELECTRIC ENGINES

As to the actual motive power, electric motors are very popular with roboteers, since they are easy to obtain and maintain, and come in various sizes. They are also very easy to control via an electronic radio-control speed controller. Most electric motors use Direct Current, and are thus powered directly from a battery.

There is an upper voltage limit for competing robots of 30 volts DC (or 50 volts AC), although the most commonly used voltages are 6, 12 and 24. A number of teams have 'over-run' their motors, running them on 24 volts instead of the intended 12 volts. This can boost the power of the robot, but does shorten the life of the motor, and brings the risk of a burn-out at a crucial moment in the competition, as more than one roboteer has discovered. A useful compromise is to design 'series connection' circuits so that a 12-volt motor runs for the most part on 12 volts, as intended, but can be switched via radio control to 24 volts for short periods, to give

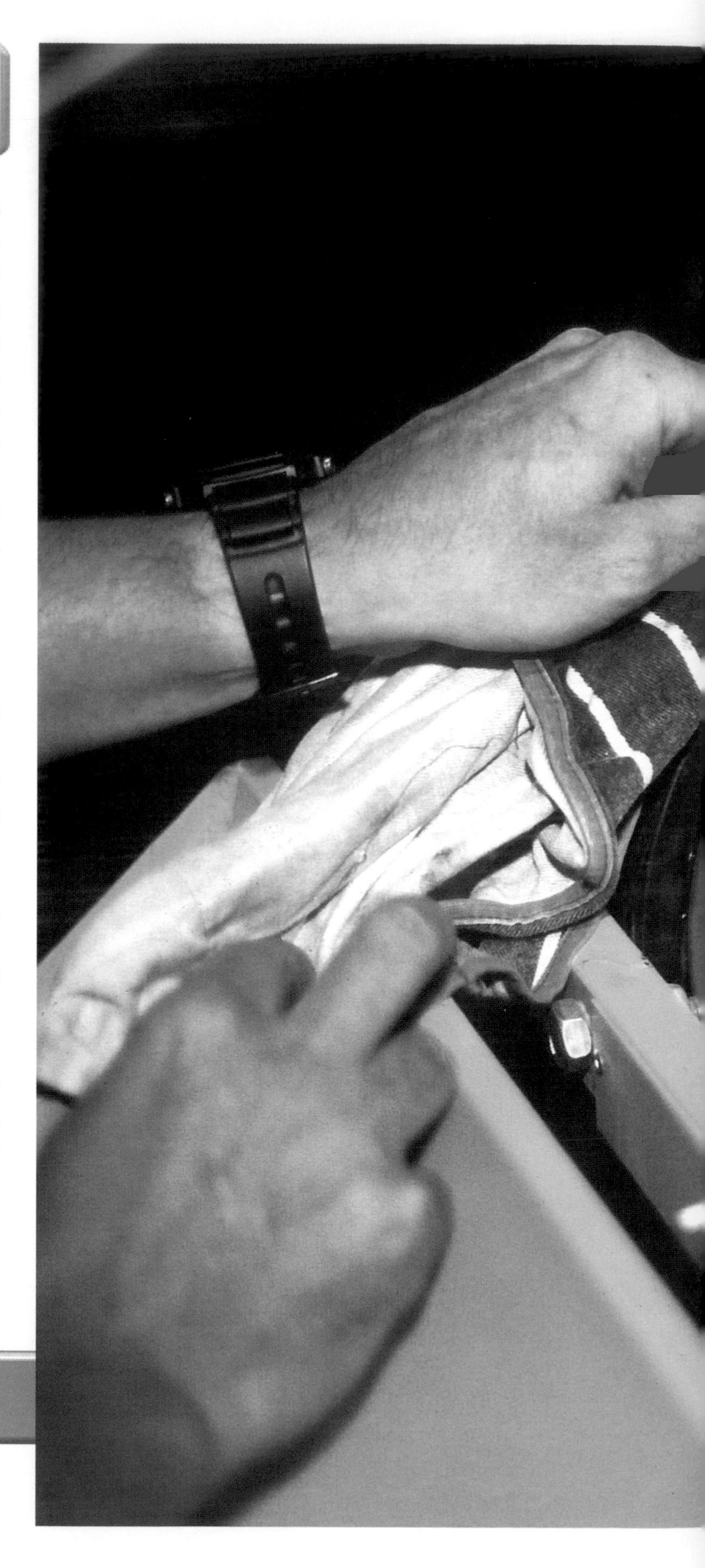

Essential maintenance to the wheel and drive chain.

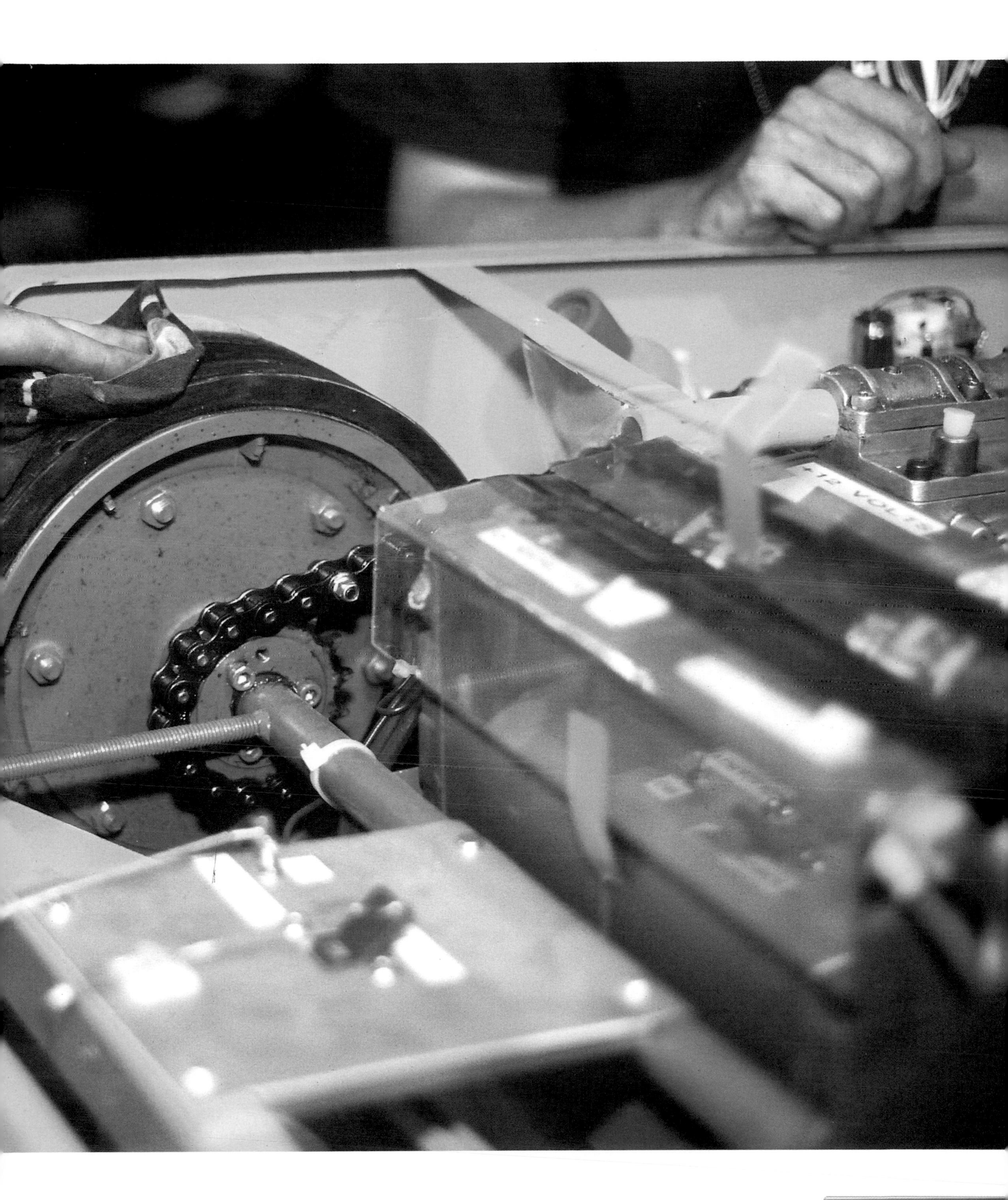
+12 VOLTS

COMPETITOR ROBOT

STING

HEAVYWEIGHT

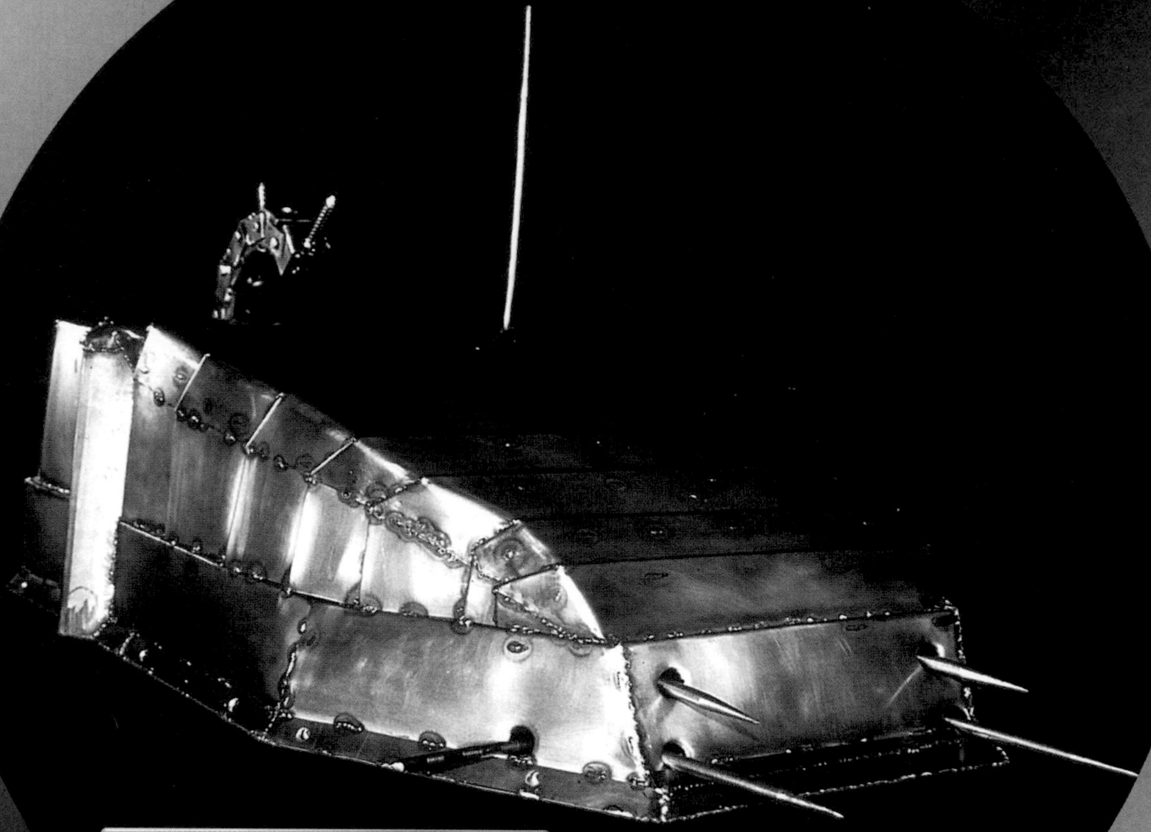

CAPTAIN: DAVID BARKER

ROBOTEERS: MATTHEW BARKER, IAN PRITCHARD

LENGTH: 160 CM

WIDTH: 70 CM

HEIGHT: 40 CM

WEIGHT: 79.3 KG

MAXIMUM SPEED: 19 KPH (12 MPH)

GROUND CLEARANCE: VARIABLE 10-40 MM

TURNING CIRCLE: ZERO

POWER: 2 X 12-V BATTERIES, RUNNING ON A 24-V SYSTEM; 3 X WHEELCHAIR MOTORS: 2 FOR TRACTION, 1 FOR WEAPON

WEAPONRY: SPIKED, JOINTED STEEL TAIL; FRONT-MOUNTED RAMMING SPIKES

that extra boost when needed. For example, you might have two 12-volt/6-amp batteries normally wired in parallel, which will give 12 volts at 12 amps. Switching to a series connection will provide 24 volts at 6 amps.

If you go for an electric engine it is very important to consider what type of battery you are going to use. This is fairly straightforward, since there are only two practical types that are easily available. The first is the Nickel-Cadmium (Ni-Cad) battery, which is rechargeable, and is standardized at 1.2 volts. The Ni-Cad comes ready-packed in specific voltages, and can be run in series for higher voltages. Although Ni-Cads have the advantage of being fast-charging, they are not particularly suitable for running large motors.

The other type is the Sealed Lead Acid Gel battery, similar to a car battery, but using a gel rather than a liquid for the electrolyte (an electrolyte is a substance that dissociates into ions in solution, thereby becoming electrically conducting). They are totally sealed, and so can be installed in any position within the robot without leaking. Although they can stand a large current drain, they cannot be fast-charged like the Ni-Cad batteries, and can take up to 24 hours to charge from flat. For this reason, it is essential for roboteers using Ni-Cads to bring at least three sets of spares to the contests.

It should be noted that standard liquid electrolyte batteries (like those used in cars) are *not* allowed. This is because if they are damaged, they will leak liquid sulphuric acid – this goes for the so-called 'sealed for life' batteries also, since, in reality, they are still not actually fully sealed.

Although both permissable battery types are low-voltage, and thus present no danger of electrocution, both are prone to rapid heating if short-circuited, and can either explode or cause fires. All circuits in a robot, therefore, must be adequately protected by thermal circuit breakers, and likewise *all* connections must be properly insulated.

INTERNAL COMBUSTION ENGINES

The other option for motive force is the internal combustion engine, which has been used by a number of competing teams, since it is more powerful than an electric one. These can be taken from go-karts, grass-cutters, small motorcycles and so on. Roboteers are advised to use four-stroke rather than two-stroke engines, since they provide more power per cubic centimetre (cc). The maximum allowed fuel capacity for internal combustion engines is 227 millilitres or six minutes' running time. This means that very large cc engines are not allowed. In addition, it should be remembered that these engines require a good deal more maintenance than electric motors.

Another victim of Sergeant Bash's flame thrower.

COMPETITOR ROBOT

CHAOS

HEAVYWEIGHT

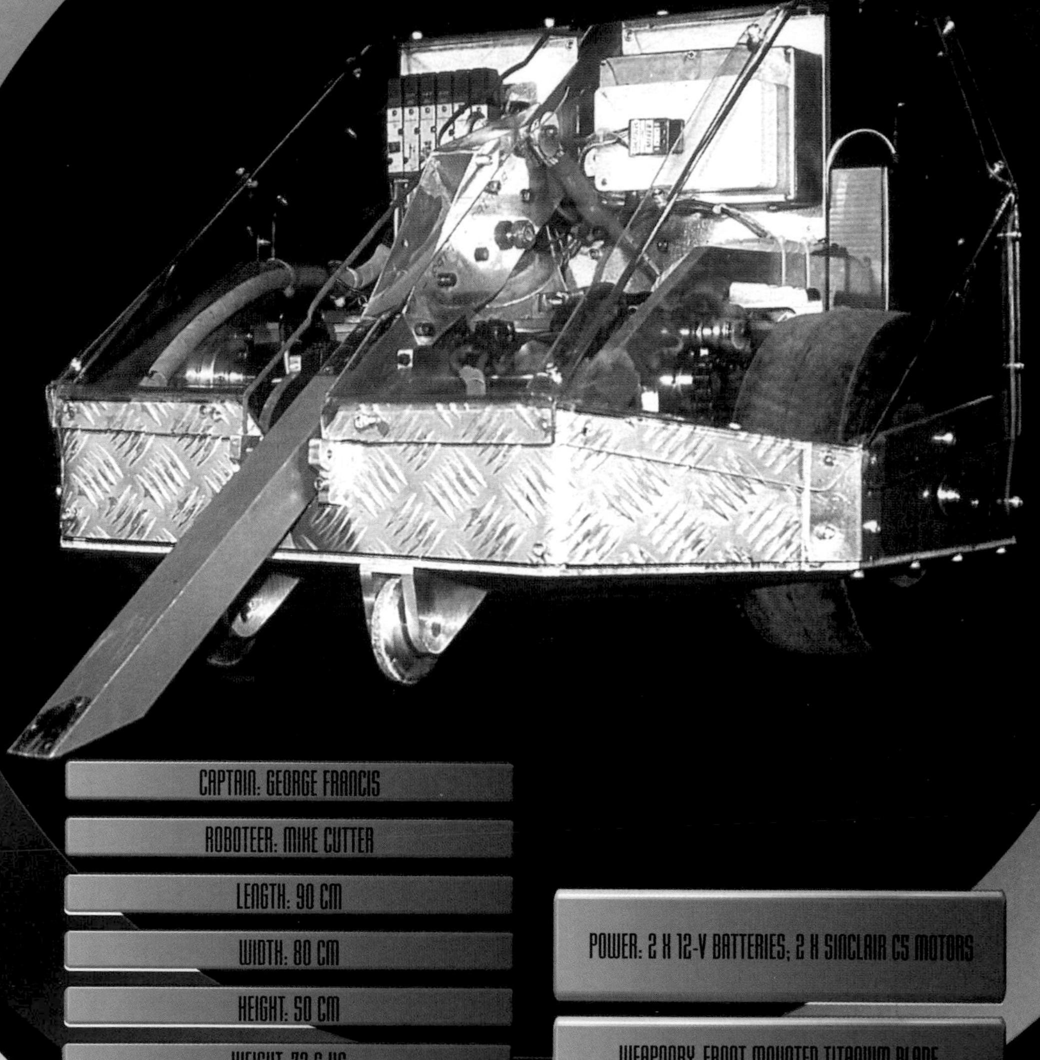

CAPTAIN: GEORGE FRANCIS

ROBOTEER: MIKE CUTTER

LENGTH: 90 CM

WIDTH: 80 CM

HEIGHT: 50 CM

WEIGHT: 73.6 KG

MAXIMUM SPEED: 32 KPH (20 MPH)

GROUND CLEARANCE: 75 MM

TURNING CIRCLE: ZERO

POWER: 2 X 12-V BATTERIES; 2 X SINCLAIR C5 MOTORS

WEAPONRY: FRONT-MOUNTED TITANIUM BLADE; FRONT-MOUNTED RAM

STEERING AND WEAPON POWER

When you have finally figured out how your robot is going to be powered, you will then be faced with another problem: how to power the steering and the weapons (if any). It is well worth considering hydraulic motors for these jobs, since they produce a great deal of power virtually instantaneously, although they do need their own power source. While many roboteers have used internal combustion engines to power their hydraulic pumps, electric motors can also be used. However, the correct British Safety Standard fittings have to be used for pipes and connectors (don't forget that hydraulics use liquid under pressure), so unless you have experience in working with hydraulics, you are probably much better off going for the internal combustion engine as a power source.

Care should be taken with all robot components which include small moving parts, since these are likely to be the first places where stress shows. In addition, when utilizing components from old or second-hand sources, it should be remembered that these may well be obsolete and difficult to replace. The Sting team, for instance, encountered quite a few problems with their finished robot, not least of which was a drive gear which

Battle of the Titans: Robo Doc and Sir Killalot go to it.

sheared during tests. Since this component was from a second-hand wheelchair, the team discovered that it was obsolete and impossible to replace. In the end, they had to get a brand-new gear cut. The first quote they received was for £160 for two new gears, which was far too expensive. Eventually, however, they found a company in Coventry who kindly did it for much less. In addition, the robot's speed controllers burned out, which necessitated running the machine on a lower power setting.

With Cassius's 12-volt motors running at 24 volts (powered by two 12-volt aircraft starter batteries), Rex Garrod, Simon West, and Ed Bull found themselves facing the constant danger of overheating. During workshop tests, the motors were run continuously for 15 minutes; they became very hot, which adversely affected the magnetic fields in the motors. The team managed to avoid this problem by running Cassius at less than full power for most of the time.

For George Francis, the most important element in designing his robot Chaos was to learn from his previous machine, Robot the Bruce, which, although extremely tough, was very hard to steer. For this reason, George opted for two main drive wheels and free-mounted castors, to increase manoeuvrability without losing speed and power. George and his teammate Mike Cutter were delighted that their robot performed well. However, George drastically underestimated the time it would take to construct Chaos. Robot the Bruce took a mere four days to build, whereas Chaos took no less than three months of spare time work. One of the main reasons for this was that the original chassis had to be rebuilt, due to the insufficient torque provided by the two motors, two drive chains and two large wheels at low speeds. The problem was solved by gearing down the robot by means of a counter shaft placed in the gearbox.

Remember also to check carefully the specifications of your source materials, particularly if they are not from off-the-shelf hobby kits. Team captain Dave Clutterbuck felt that his robot Demon's main weakness was its lack of speed, and that this was due to the motor having come from an indoor wheelchair, which was less powerful than an outdoor model.

It is also extremely important to make sure that the motors powering your robot's drive wheels are running at equal speeds. While this might seem an obvious statement, it can easily be overlooked, and even a minor difference in running speeds can make your machine very difficult to steer. This is what happened to Richard Peter with his robot Flirty Skirty, which developed an unfortunate tendency to drive in a circle because the motors, taken from Volvo windscreen wipers, were running at slightly different speeds.

However powerful, sometimes the robots just can't get away in time.

COMPETITOR ROBOT

CASSIUS

HEAVYWEIGHT

CAPTAIN: REX GARROD

ROBOTEERS, SIMON WEST, EDWARD BULL

LENGTH: 145 CM

WIDTH: 84 CM

HEIGHT: 35 CM

WEIGHT: 79.3 KG

MAXIMUM SPEED: 40 KPH (25 MPH)

GROUND CLEARANCE: VARIABLE 0–110 MM

TURNING CIRCLE: ZERO

POWER: 2 X 12-V AIRCRAFT STARTER BATTERIES

WEAPONRY: CO_2 GAS-POWERED RAM

TRANSMISSION

Transmission components include steering, gearboxes, drive-belts, suspension, and wheels, each of which requires considerable thought, since the failure of any robot part can result in instant defeat for the team.

STEERING

The steering mechanism can be of the rack and pinion type used by cars and other road vehicles – that is, with either the front or rear wheels controlling direction – or it can be the tank-type – with independent motors controlling each side of the robot. Car-type steering can be a risky option, since the steering mechanism will be placed under a great deal of strain, with a lot of wear and tear on individual components. Tank-type steering is altogether more sturdy, and can give much greater manoeuvrability. As its name suggests, this method of steering operates like a tank, in which power is disengaged from the left-hand track to turn left, and the right-hand track to turn right. However, this principle can also be applied to wheeled robots. Typically, the chassis contains two wheels mounted at the central point along the forward axis, with freely rotating (unpowered) castors at front and rear to provide stability. Although a single motor can be used to drive each wheel through an individual transmission, a more effective option is to have a separate, individually-controlled motor driving its own wheel. This option lends greater speed to each manoeuvre: when one motor is operating forward and one in reverse, the robot will turn much faster than with one wheel merely disengaged. With this arrangement, the robot can also spin on the spot – a manoeuvre much loved by audiences following a victory.

GEARS

If your robot's wheels are not directly driven by the motor (this requires quite a lot of power), you will need to include some form of 'power chain' – a means of getting the power from the motor to the drive wheels. This can be achieved either through a simple gear mechanism, or by means of belts and pulleys. It is essential to choose the largest gears and the strongest belts possible, since these components will be placed under a great deal of strain, and they are invariably the weakest points in the power chain.

A hammer can work wonders on a recalcitrant robot.

SUSPENSION

An adequate form of suspension is also well worth considering, in view of the harsh treatment experienced by competitors during *Robot Wars*. It is important to remember that, even when you smash into an opponent's robot, your machine will also receive a nasty jolt from the impact of the collision. A good suspension system can also prevent wheelspin when applying sudden power to the drive wheels.

No time to relax, as long as your robot is in with a fighting chance.

COMPETITOR ROBOT

GRIFFON

HEAVYWEIGHT

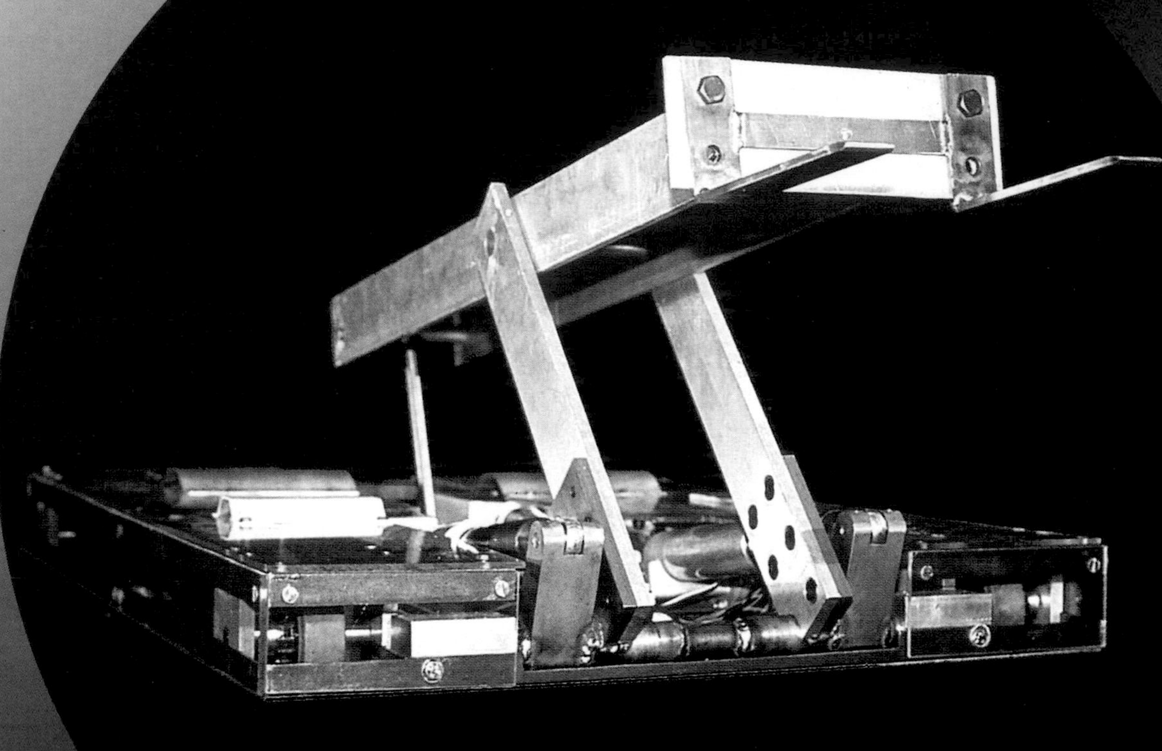

CAPTAIN: OLIVER STEEPLES

ROBOTEER: BEN STEEPLES

LENGTH: 130 CM

WIDTH: 70 CM

HEIGHT: 20 CM

WEIGHT: 79.4 KG

MAXIMUM SPEED: 29 KPH (18.5 MPH)

GROUND CLEARANCE: 13 MM

TURNING CIRCLE: SHOULD BE ZERO, BUT 1 METRE IN PRACTICE

POWER: 2 X 12-V BATTERIES; 750-WATT BOSCH MOTOR.

WEAPONRY: LIFTING ARM

Elvis has a problem spotting the ball.

WHEELS

As for the wheels themselves, these can either be solid, or have pneumatic tyres. However, pneumatic tyres are preferable since, while vulnerable to punctures from a well-aimed spike or cutting implement, they do provide additional grip, as well as partial suspension. It is also advisable to make sure that you attach the tyres to the wheels as securely as possible, since high torque can cause a wheel to spin inside its tyre.

Neil Lambeth and John Ebdon of the Elvis team managed to convert their design to a working machine for the First Wars without too many problems – apart from time and money (factors mentioned by a number of roboteers). However, their problems increased somewhat during the Second Wars. They used pretty much the same design for Elvis as they had used in the First Wars, but when they took it to London for the initial checks, they discovered that Elvis was a little overweight. Back in the workshop, Neil and John ended up rebuilding the entire chassis, a job which they only finished the day before filming started. Not having a great deal of time to test their redesigned machine, they drove it around the car park outside the studio a few times.

It was then that they discovered another problem: one of the drive belts connecting the motors to the wheels was slightly looser than the other, causing the robot to drift to one side when moving forward. This could be compensated for with careful steering, but it was easy to oversteer while trying to compensate, so controlling the robot was rather difficult.

All this resulted in a rather embarrassing moment for the Elvis team during the Soccer game in the Trials. Elvis and one other robot were left on the field (the other robots had all scored, and consequently had been removed). When the other robot broke down, the house robots instantly went in for the kill, leaving an open goal for Elvis. At this point his steering started giving problems again, and Elvis could not take advantage of this golden opportunity. This caused no end of amusement among the studio audience, who started a slow hand-clap. However, an exciting deflection from the goalkeeper eventually gave Elvis the necessary goal.

Once their robot Griffon had been built, tested and dispatched to the Wars, Oliver Steeples and his brother Ben encountered a problem with the wooden floor panels of the Arena, which caused too much friction against the wheels and inhibited steering. The Griffon team also had some problems with interference. Oliver later added screws to Griffon's wheels, which improved the robot's steering ability considerably. He also replaced the original electronic speed controllers with mechanical ones, to avoid the interference problems.

In common with a number of other roboteers, Simon Harrison, Phil Brett, and Steve Monk of the King Buxton team had serious problems both with time and money. They decided to go for the simplest construction possible: a box section chassis, solid axles and solid rubber wheels, with many structural components welded together to reduce maintenance on the robot. The main problem the team had during the Wars themselves was with electrical components heating up. When King Buxton was run for longer than a minute, the danger of the motors heating up increased, especially since the insulation in the motors was progressively burned away. In addition, the motors passed more current than had originally been thought, and when this happened, electrical resistance increased and the power to the drive wheels consequently fell. The team tried to solve this problem by installing fans in the robot, but this was not very effective, since everything was so tightly packed inside the shell. As if this were not enough, the solid axles proved incapable of taking any serious punishment, and started to buckle under the pressure of the knocks the robot was receiving.

Added to this was an additional problem with the microprocessor-controlled steering, which lost its zero setting, with the result that King Buxton developed a habit of drifting off to the right during the Gauntlet. Simon also had a problem with interference from the spark plugs in the motor powering house robot Matilda's chainsaw. Every time Matilda's chainsaw started, King Buxton jumped forward!

David and Goliath: the mighty Robo Doc has some trouble with King Buxton.

COMPETITOR ROBOT

KING BUXTON

HEAVYWEIGHT

CAPTAIN: SIMON HARRISON

ROBOTEERS: STEVE MONK, PHIL BRETT

LENGTH: 90 CM

WIDTH: 85 CM

HEIGHT: 27 CM

WEIGHT: 78.0 KG

MAXIMUM SPEED: 29 KPH (18 MPH)

GROUND CLEARANCE: 50 MM

TURNING CIRCLE: ZERO

POWER: 4-WHEEL DRIVE

WEAPONRY: TITANIUM SPIKES

RADIO CONTROL

This is another vital aspect of robot design. To begin with, you will need to consider how many channels you require; the minimum is two, one for drive and one for steering, but if you have any other features on your robot, such as weapons, you will need more channels. It is important that the channels controlling drive and steering are 'proportional', which means that speed and steering position can be changed gradually. Most control sets will include at least two proportional channels.

WHICH FREQUENCY?

In the UK, four frequency bands are allowed for amateur radio control by the Home Office: 27MHz, 35MHz, 40MHz and 459MHz. However, of these four frequencies, it is most likely that you will only be allowed to use 40MHz. This is because 35MHz is restricted to flying models only, the house robots use the 459MHz frequency, and CB radio uses 27MHz.

WHERE TO PUT YOUR RECEIVER

You should think carefully about where to position the radio receiver on your robot. For instance, it should be placed as far away as possible from the electric motor, or the spark plugs of internal combustion engines. Although it is never a good idea to cut the receiver aerial, it does not necessarily have to be stretched out at full length: it can be wound up and taped inside the robot's body without affecting its performance. However, all-metal robot bodies can create a phenomenon known as a Faraday Cage, which inhibits electrical signals, so you are advised to experiment with the position of your receiver and its aerial.

And remember that there is no rule against attacking an opponent's aerial and trying to snap it off!

SPEED CONTROLLERS

It is advisable to use electronic speed controllers rather than the switched, 'stepped' type (that is, with slow, medium, and fast settings) as these components are usually the first to fail. This is because when an electric motor stalls, as can happen when robots collide, the current going through the speed controller rises dramatically, and when a motor reverses direction, the current can peak at 1,000 amps or more, which burns out the component. Easily available off-the-shelf speed controllers have peak ratings of several hundred amps; however, this only applies to electronics running at low temperatures, which rarely happens in *Robot Wars*. It is worth remembering that, of all the off-the-shelf technology that has been used by roboteers, the most successful has been from wheelchairs.

THE ELECTRONIC FAIL SAFE

Electronic fail safes are essential, and no robot will be allowed to compete unless it is fitted with one. A fail safe will prevent the operation of a radio control switch or servo, should the signal from a transmitter fail or be degraded. It is also likely, particularly for the heavier classes, that fail safes will also be required on the drives.

INTERFERENCE

Some of the robots in *Robot Wars*, and this includes the house robots, have suffered problems with interference. This can be caused by a whole range of things – whether you are using AM or FM radio-control sets, damage caused to a receiver aerial during combat, the fact that the drivers' booth is constructed from scaffolding, and so on. Competitors in *Robot Wars* are encouraged to think carefully about this phenomenon – after all, you don't want your machine to fail at a crucial moment in the conflict.

Ozzie Boulter's robot Wizard included an interesting design innovation. The machine contained four servos, one of which operated the fail safe mechanism (a servo is an electric motor or hydraulic piston that supplies

COMPETITOR ROBOT

ELVIS

HEAVYWEIGHT

CAPTAIN: NEIL LAMBETH

ROBOTEER: JOHN EDBON

LENGTH: 140 CM

WIDTH: 145 CM

HEIGHT: 80 CM

WEIGHT: 83.0 KG

MAXIMUM SPEED: 16 KPH (10 MPH)

GROUND CLEARANCE: VARIABLE 10-140 MM

TURNING CIRCLE: ZERO

POWER: 2 X 12-V/24-V (7-AMP) BATTERIES; 2 X 12-V WHEELCHAIR MOTORS

WEAPONRY: REAR-MOUNTED 30-CM DISC CUTTER; FRONT-MOUNTED LIFTING MECHANISM

COMPETITOR ROBOT

WIZARD

HEAVYWEIGHT

CAPTAIN: OZZIE BOULTER

ROBOTEERS: RICHARD JOHNSON, SUSAN DARBY

DIAMETER: 75 CM

WEIGHT: 79.4 KG

MAXIMUM SPEED: 24 KPH (15 MPH)

GROUND CLEARANCE: 3–4 MM

TURNING CIRCLE: ZERO

POWER: 2 X 12-V BATTERIES (24-V SYSTEM); AMERICAN MOTORS, ORIGIN UNKNOWN

WEAPONRY: TWO SIDE-MOUNTED CHAIN WHIPS; TWO SIDE-MOUNTED DRILL BITS

In the spotlight: a few seconds of intense concentration.

Wizard receives some much needed attention.

power to a moving device, or servomechanism). Normally, one servo operates one device, but Ozzie figured out a method of operating more than one device with a single servo. He did this by making a series of thin metal strips, which he then superglued to a copper paddle, and connected the paddle to several different moving features. In this way, Ozzie's three remaining servos could be made to operate up to 12 separate devices, thus saving a considerable amount of space inside the robot. A number of other teams commented on the ingenuity of this idea.

The worst moment for the Killertron team, Ian and Abdul Degia, and Richard Broad, came when their robot began to suffer from interference from an unknown source. This resulted in the servos and its pickaxe weapon operating by themselves. In common with several others who participated in *Robot Wars*, Abdul felt that UHF transmitters would solve these interference problems. Robo Doc team captain Mike Franklin felt that the biggest problem faced by the team during *Robot Wars* was mechanical breakdown, and Robo Doc suffered badly from this during the initial parade due to interference. He agrees with Abdul that the interference problem could be solved through the use of UHF transmitters.

COMPETITOR ROBOT

NAPALM

HEAVYWEIGHT

CAPTAIN: DAVID CROSBY

ROBOTEERS: CLAIRE GREENAWAY, REBECCA GLENN

LENGTH: 200 CM

WIDTH: 100 CM

HEIGHT: 125 CM

WEIGHT: 84.1 KG

MAXIMUM SPEED: 24 KPH (15 MPH)

GROUND CLEARANCE 5 MM

TURNING CIRCLE: ZERO

POWER: 2 X 12-V/24-V BATTERIES; 3 MOTORS: 2 X DRIVER MOTORS (WHEELCHAIR-DERIVED), ADDITIONAL MOTOR TO POWER WEAPONRY

WEAPONRY: TITANIUM-FRONTED RAM/SCOOP; REAR-MOUNTED CHAINSAW WITH 40-CM BLADE; REAR-MOUNTED STAINLESS STEEL PIERCING ARM

STYLING

Away from the tight disciplines of mechanics and electronics, styling is the aspect of robot design where competitors can really let their imaginations take over. Since they are usually confined to the exterior body shell of a robot, styling considerations are usually centred on strength and durability. After all, it is not much use having an elaborately designed outer shell that is torn to pieces in the first five seconds of combat! But there are other factors to consider. Firstly, while wedge- and box-shaped robots have proved successful in the past, they are not particularly interesting visually, and as *Robot Wars* is primarily a televised event in Britain the producers are very keen to see designs that display imaginative flair. So you will greatly increase your chances of competing if you have a visually interesting machine. Secondly, the Judges are always on the lookout for nicely styled machines, and even if you don't become a champion, you may well pick up the Best Design Award.

Claire Greenaway, Becci Glenn, and their teacher David Crosby of Dartford Girls' Grammar School, wanted to base the design for their robot Napalm around some sort of animated feature. They settled for an imaginative system in which an armature lifted up to reveal a chainsaw, ready to go in for the kill. However, the chainsaw was snipped off during combat before it could be used, and they couldn't fix it between bouts. Nevertheless, the moving armature with its twin metal spikes remained a visually interesting feature, as well as a highly effective weapon.

The shape of Napalm's body shell evolved from a number of prototype designs, which included a ball-shaped front end with a tail, and a paint-job reminiscent of a flame. They finally decided on an arrowhead shape, although the flame concept was retained, and inspired the robot's name.

David and the girls wanted a robot that was visually interesting, something that had a personality and character of its own. They felt that the best way to achieve this was through the inclusion of eyes, which are inherently expressive features. The Napalm team had noted during their participation in the First Wars that robots such as Matilda and a competitor robot named Nemesis had eyes, and this alone was enough to make them stand out from the crowd. However, painted eyes were not enough for David and the girls. Instead, they went for an even more dynamic look, and installed halogen light bulbs in two huge sockets. The result was Napalm, one of the most unusual and exciting designs seen in the Second Wars.

For Neil Lambeth and John Ebdon, the most important aspect of their robot Elvis's initial design was fun. The one thing they didn't want was a Mad Max-type machine, they wanted something a little more lighthearted. They struck on the idea of a UFO motif, which originally was to have had an alien driver in the cockpit. However, Neil and John were still not convinced. They eventually had the brilliant notion of having Elvis Presley as the UFO pilot! (You may be aware of the theory that Elvis did not die, but went to live on another planet.) Apart from presenting a splendidly bizarre sight, the conical design of Elvis has a practical application which is worth bearing in mind: not only is a curved surface hard to grab and hold, it also deflects blows very nicely.

Sting is perhaps unique in *Robot Wars*, in that its design was directly inspired by an animal – to be precise by David Barker's pet cat Chloe. The cat has a deformed tail, which is permanently curved over its back. David's son Matthew decided that this would be a good idea for a feature on their robot. David worked on the mechanics of the tail. The original concept was for a prehensile tail, complete with controlling ligaments. A cable would run down through the segmented tail, mimicking the action of a tail in the animal kingdom.

However, the team soon found that the forces placed upon the tension spring needed to retract the tail were too great to be practical, and so they went for a much simpler crank-operated system to curve the tail back.

AHEAD
CLOSED

Roadblock is put through its paces.

This required a little less assistance from the tension spring, and although the tail in the finished robot was more rigid and curved back a little less than the team had intended, it was still functional, and remains one of the most original and innovative features seen in *Robot Wars* so far.

Cassius is a particularly sleek, stylish and well designed robot, and benefits considerably from builder Rex Garrod's years of experience as a special-effects designer. The initial design evolved from the team's previous robot, Recyclopse, which saw action in the First Wars. Many of Recyclopse's components, such as motors, valves and so on, found their way into Cassius. In fact, according to Rex, when building Cassius they placed loads of Recyclopse's bits on the floor of his workshop and then simply fitted them together. Of course, the design process was actually somewhat better thought out than that, and they drew up the initial design on a computer.

The chassis was constructed of the cheapest steel tubing the team could find, while the CO_2-powered ram, which operates at a pressure of approximately 737 kilograms per square centimetre and is capable of lifting 1.5 tonnes, is mounted in a racing-car roll cage. The ram was designed to reduce pressure automatically three-quarters of the way through its arc, otherwise it would have destroyed itself on contact with a target. It is actually very important for those intending to include powerful moving features in their robot to remember that, on contact with an enemy, a shock wave will travel back into your own machine as well as into your opponent. You must make absolutely certain that the structure of your robot will be able to withstand the force of these physical shocks.

For Rex one particularly useful and satisfying aspect of his robot Cassius is that it can be used for educational purposes. Rex regularly lectures to schools on the subject of engineering and robot technology. He is also a marvellous sportsman, and at first was seriously concerned that he was somehow cheating in *Robot Wars*, due to his experience in special effects. This, of course, couldn't be further from the truth. Competitors are perfectly at liberty to bring whatever experience they may have to the design and construction of their robots. After all, this is the most challenging competition in the roboteers' world.

In designing Wizard, Alwyn 'Ozzie' Boulter's primary concern was for his robot to stay out of trouble. For that reason, he first considered a much-opted-for wedge- or dome-shape, before finally settling on a squat, low-profile cylinder-shape. Ozzie considered this to be the optimum design to avoid getting caught by or wedged into obstacles. However, on consideration he decided that the cake tin-like shape his robot developed was too plain and uninteresting. Second thoughts gave him the idea of a wizard's hat-shape. When asked where his inspiration for this idea came from, Ozzie replied that it might have had something to do with his nickname and a certain famous wizard – from Oz!

Wizard's chassis is made from a composite of wood, plastic, and metal, while its outer shell is a composite of plastic and fabric. As an added touch, the upper hat section was fitted with flashing lights. During the Wars, Ozzie had problems with the robot's low ground clearance of 3 to 4 millimetres. The robot was also rather difficult to steer, as Ozzie found out when he conducted a 15-second test drive half an hour before going to London to participate in the Wars, during which Wizard managed to smash just about every breakable object in his conservatory. In fact, these steering difficulties made Wizard one of the most unpredictable robots. No one knew where it was going to go next, mainly because Ozzie himself didn't know! In the initial parade, Wizard went everywhere but where it was supposed to go, and it took five takes just to get the robot on to the floor of the Arena.

Wizard's various problems continued during the Wars themselves. At one point, the drive chain slipped, became jammed underneath it and lifted it off the floor. To add to his problems Wizard's weapons – two chain whips and two drill bits bolted to the inside of the cylindrical body – did not work particularly well against his opponents.

Twelve-year-old Dominic Rott of the Rottweiler team wanted to base his design for their robot on one of the machines from the computer game *Red Alert*. They ended up with a powerful, robust, flame-proof machine, armed with vicious 15-centimetre steel spikes around its perimeter. Its shell is fashioned from 2-millimetre steel. The robot's profile is virtually a triangular prism, constructed of featureless, slate-coloured steel, which gives Rottweiler a subtly menacing appearance. Apart from the usual problems presented by a shortage of time for design and construction, Dominic's team also suffered a serious setback just before they were due to take their robot to London for the viewing weekend when robots are selected for the television series, when all the circuits shorted and the internal wiring burned to a crisp. Fortunately, they managed to solve this rather demoralizing problem just in the nick of time to compete.

One of the most visually striking robots to take part in both the First and Second Wars is Nemesis, whose most immediately noticeable feature is its fur-covered body (black polka dots on a red background). Added to this are a fiendish, toothy grin, and two huge, bloodshot eyes mounted on its wheel arches.

There is no doubt in the minds of anyone involved in *Robot Wars* that the variety of styling shown by the machines is one of the most astonishing things about the whole phenomenon of the Wars. It is also everyone's wish that this should continue to the maximum.

COMPETITOR ROBOT

NEMESIS

HEAVYWEIGHT

CAPTAIN: PETER RAYMOND

ROBOTEERS: JOE GAVIN, WILLIAM MURPHY, JOHN CUNNANE

LENGTH: 130 CM

WIDTH: 123 CM

HEIGHT: 55 CM

WEIGHT: 74.5 KG

MAXIMUM SPEED: 9 KPH (6 MPH)

GROUND CLEARANCE: 50 MM

TURNING CIRCLE: ZERO

POWER: 2 X 12-V BATTERIES; 1 X 6-V BATTERY; 2 X 24-V MOTORS

WEAPONRY: REAR-MOUNTED SWORD BLADES; FRONT-MOUNTED CUTTING BLADES

BODY CONSTRUCTION

The body of your robot is no less important than the chassis, since not only will it have to take the brunt of the brutal action, it also provides the best opportunity to display your ingenuity to the audience and, of course, the Judges.

The first challenge you will have to face is how to secure the body firmly to the chassis (you won't want to see it flying off in the middle of a battle), while making sure that it can be easily removed in the very likely event that maintenance and repairs are required in the pit area.

MATERIALS

Although the word 'robot' calls to mind a metallic creature, it is well worth considering materials other than metal for its body. Wood, for instance, although decidedly 'low-tech', is very resilient to impact, and can even jam saws trying to cut through it. The same goes for fibreglass and other plastic composites (don't forget that the fearsome Matilda's body is made of fibreglass). Although circular saws can occasionally present problems, a fibreglass body moulded into a curving, organic shape can be very difficult to grab hold of, and thus may well elude such weapons. And remember that an interesting body design will impress the judges.

The best of enemies: Napalm and Sir Killalot.

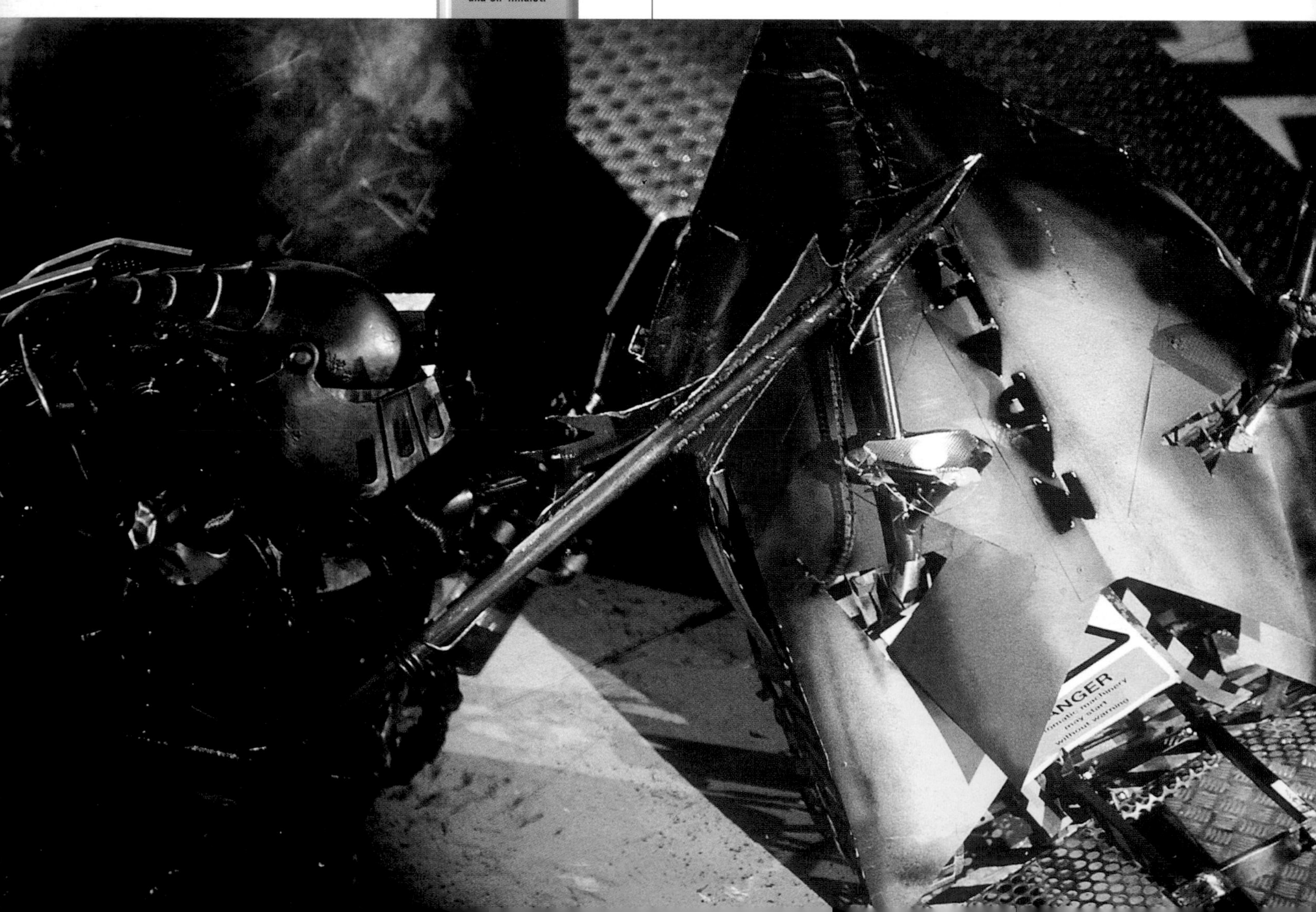

Ever get the feeling you're being demolished? Sir Hillalot in action.

The particular strength of Napalm, designed and built by David Crosby, Claire Greenaway, and Becci Glenn, is its extremely tough and durable titanium hull, which not only could stand up to severe knocks and attacks from other robots, but also had the feel of a house robot (with which the team was very impressed in the First Wars).

Sting, built by David Barker, his son Matthew, and Ian Pritchard, has several design advantages, but the fact that it is extremely strong – it's built around a steel ladder chassis – is its key element. The 7-millimetre-thick mild steel armour shell is double-skinned in several places, to provide additional structural strength.

Mike Rickard made the shell of his robot GBH with an underskin of perforated steel mesh overlaid with three layers of fibreglass, resulting in a very smooth finish which proved effective in deflecting the weapons of several fierce opponents. In fact, one or two other contestants told the GBH team that a fibreglass body was unsuitable to the rough-and-tumble of *Robot Wars*, and wouldn't last very long. In the event, the very opposite proved to be true. GBH stood up to punishment very well indeed.

Peter Duncanson explained that polypropylene, which he used for the body of his robot Spin Doctor, is an extremely hard plastic (it is apparently the material from which shower cubicles on oil rigs are made); it also has excellent non-stick qualities (the next step down from Teflon), and thus is very difficult to grip.

Another innovation was seen on Milly-Ann-Bug. She was coated with something called kevlar. This is very durable (it is one of the components used in bullet-proof vests), and was also useful in gumming up other robots' chainsaws.

WEAPONRY AND FIGHTING ABILITY

Given the extremely violent nature of *Robot Wars*, it should go without saying that the machine with the most effective weapons has the best chance of securing victory. Not only will you win over the Judges, but also the studio audience and the people watching at home. However, as with any sporting event, safety considerations are paramount, and for this reason certain weapons systems have to be banned. These include electronic weaponry, such as lasers and other stun guns; liquid weaponry, such as water, glue and expandable foam; non-tethered missiles and other projectiles; and deliberate electronic interference.

However, what may be considered a restriction by some can be considered a challenge by others. For instance, you can include rotating weapons such as chainsaws, circular cutting discs or spinning platforms with spikes or blades attached; or concussive weapons such as hammers and mallets; then again, why not go for pincers to immobilize an opponent, or a lifting device to overturn them.

Another strategy worth considering is to have interchangeable weapons or other body components on your robot, as the Plunderbird team did with their machine. While this falls within the rules, you will not be allowed to have so many different components that you are, in effect, entering two different robots: the basic chassis and body must remain the same. You may also decide to exchange one component for another to meet a particular challenge, for instance adding some form of scoop for the Soccer game. Remember, however, that time is always of the essence, and you will not have long to alter the configuration of your machine in the pits.

Razer, built by Ian Lewis and Simon Scott, was one of the most aggressive-looking designs in the Second Wars. According to team captain Ian Lewis, the main attribute he and Simon Scott were looking for was the most effective method of destroying an opponent. Ian had noted with disappointment that during the First Wars no robots were completely destroyed. He wanted something that could obliterate an opponent, rather than just putting a hole in it. For him and Simon, the only answer was hydraulics. They had considered spikes, hammers, grinders, and even explosives (which of course are not allowed). But they eventually settled for a long arm, mounted towards the rear of the robot, with a pointed tip which snaps down, giving the entire machine the appearance of an enormous lobster claw.

Ian is another roboteer who has grappled with the problem of leverage. When an arm crashes down on an opponent, part of the kinetic energy is unavoidably transmitted back into the attacking robot. The most common result of this is that the machine is levered up off the floor, and the energy that could have been directed at the opponent is wasted. The solution arrived at by the Razer team was as simple as it was brilliant. Instead of mounting the weapon arm at the front of the robot's body, they attached it to the rear of the chassis, while choosing a wedge-shaped profile for the body. In this way, Razer can slide underneath an opponent, and then bring the hydraulically powered, sharp-pointed arm down, thus using the entire robot as a grabbing mechanism. The kinetic energy that would have been wasted is used to bring Razer's body up under its victim and increase the force of its grip.

For Shane Howard and Brian Fountain of the Mace team, there were problems during the *Robot Wars* themselves. Although Mace's lifting arm was very powerful, it was also very slow. So whenever it slid beneath an opponent's robot, the machine had time to slide off it before being overturned. The rotating flail fared slightly better (being able to work at any angle), but Shane and Brian found that it still wasn't quite as effective as a straightforward disc-cutter.

In designing Griffon, Oliver Steeples' main concern was with subtlety as opposed to naked aggression. For this reason, he decided to avoid all the weapons usually associated with *Robot Wars*, such as chainsaws and hydraulic axes. He said that while they certainly look

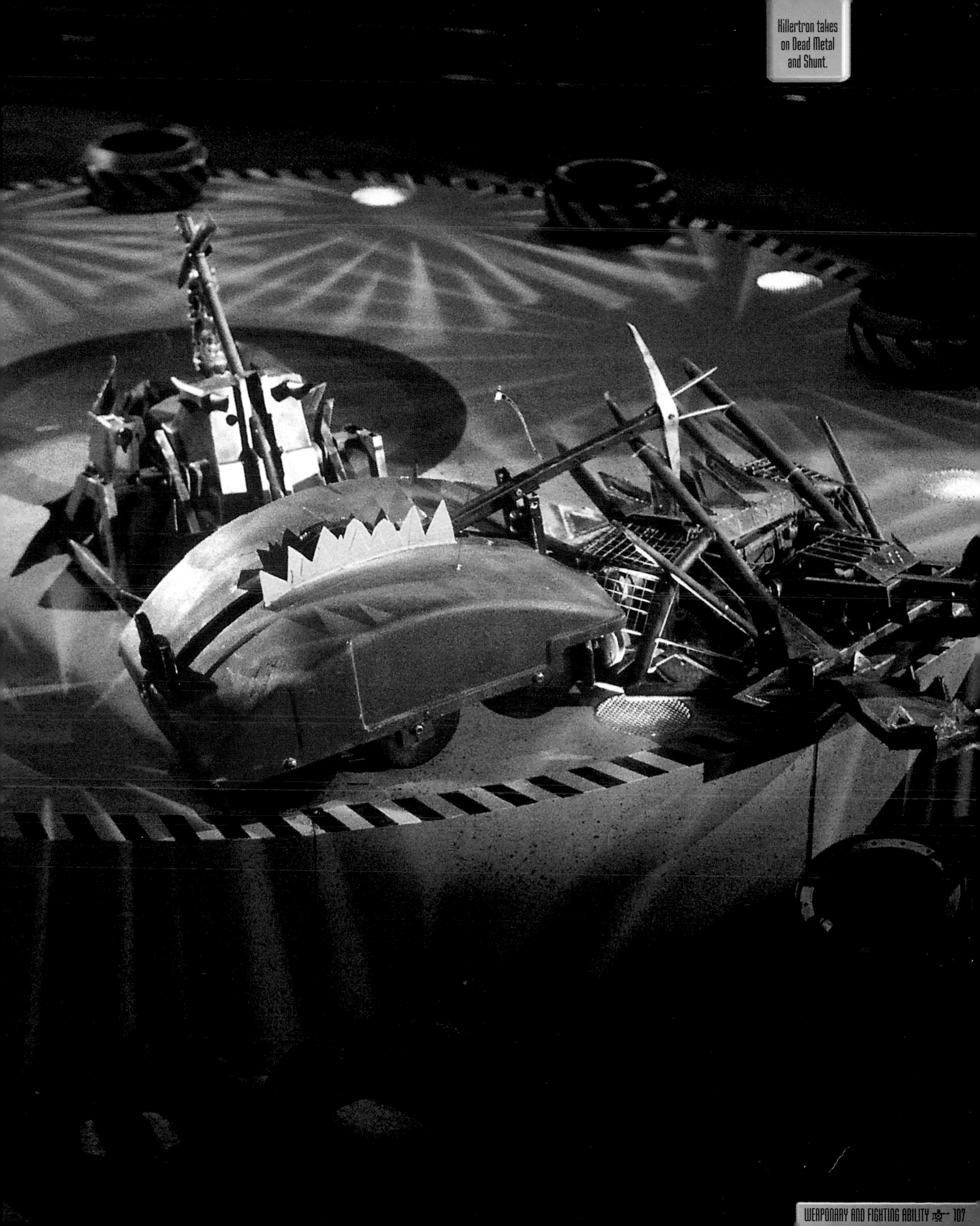

Killertron takes on Dead Metal and Shunt.

COMPETITOR ROBOT

MACE

HEAVYWEIGHT

CAPTAIN: SHANE HOWARD

ROBOTEER: BRIAN FOUNTAIN

LENGTH: 130 CM

WIDTH: 80 CM

HEIGHT: 40 CM

WEIGHT: 79.4 KG

MAXIMUM SPEED: 9 KPH (6 MPH)

GROUND CLEARANCE: 15 MM

TURNING CIRCLE: ZERO

POWER: 2 X 12-AMP BATTERIES;
2 X MOTORS WITH 2-WHEEL DRIVE

WEAPONRY: FRONT-MOUNTED RAM/LIFTER;
REAR-MOUNTED SPINNING FLAIL WITH 4 BLADES

good when the sparks fly, they actually don't do a great deal of damage, particularly if an opponent has a well-designed and well-armoured body shell. He decided that the best weapon was actually a lifting arm, since an overturned robot is very unlikely to be able to right itself, unless it has a feature specifically designed for that purpose. So Griffon was designed entirely around its central lifting arm, which gives the robot a flat, featureless and deceptively unthreatening appearance, but which is nevertheless able to lift 86 kilograms. Due to his keen interest in engineering, Oliver was able to work out all the stresses and tolerances beforehand; there was no guesswork involved.

Cassius is one of the very few robots with a self-righting mechanism. In fact, nobody involved with *Robot Wars* had any idea of Cassius's capability, until it was overturned, and team captain Rex Garrod proceeded to activate the gas-powered ram. Cassius instantly sprang into the air, and landed back on its wheels, much to everyone's astonishment.

According to Mike Rickard, GBH's main weakness was that the drive belts had a habit of slipping on the pulleys. The robot's strong points were its powerful pneumatic flipper, and its steel and fibreglass bodywork.

Robin Herrick felt that Bodyhammer's main strengths were its speed and manoeuvrability, and also its strong, conical shell, which can easily deflect blows. Its main weakness was a very low ground clearance, which can cause problems when trying to negotiate ramps. It should be remembered that weaponry can be defensive as well as offensive, and a cunningly designed body shell can serve a robot well in tight situations. Bodyhammer's shell design is a case in point. The truncated cone presents two surfaces that are, by their very nature, good at deflecting attacks: the curved flanks, and a flat but steeply-angled front surface.

George Francis saw the flipping mechanism on Recyclopse, and thought it was such a good idea that he decided to incorporate something similar in his robot Chaos. Unfortunately, however, it was not as successful as it might have been, due to the gas valve being too small, and the Sodastream bottles powering it not having enough volume. The result was that the lifting ram was not fast enough when a load was placed on it. Also, rather ironically, Chaos ended up being harder to

Nemesis and Onslaught battle it out.

COMPETITOR ROBOT

LEIGHVIATHAN

HEAVYWEIGHT

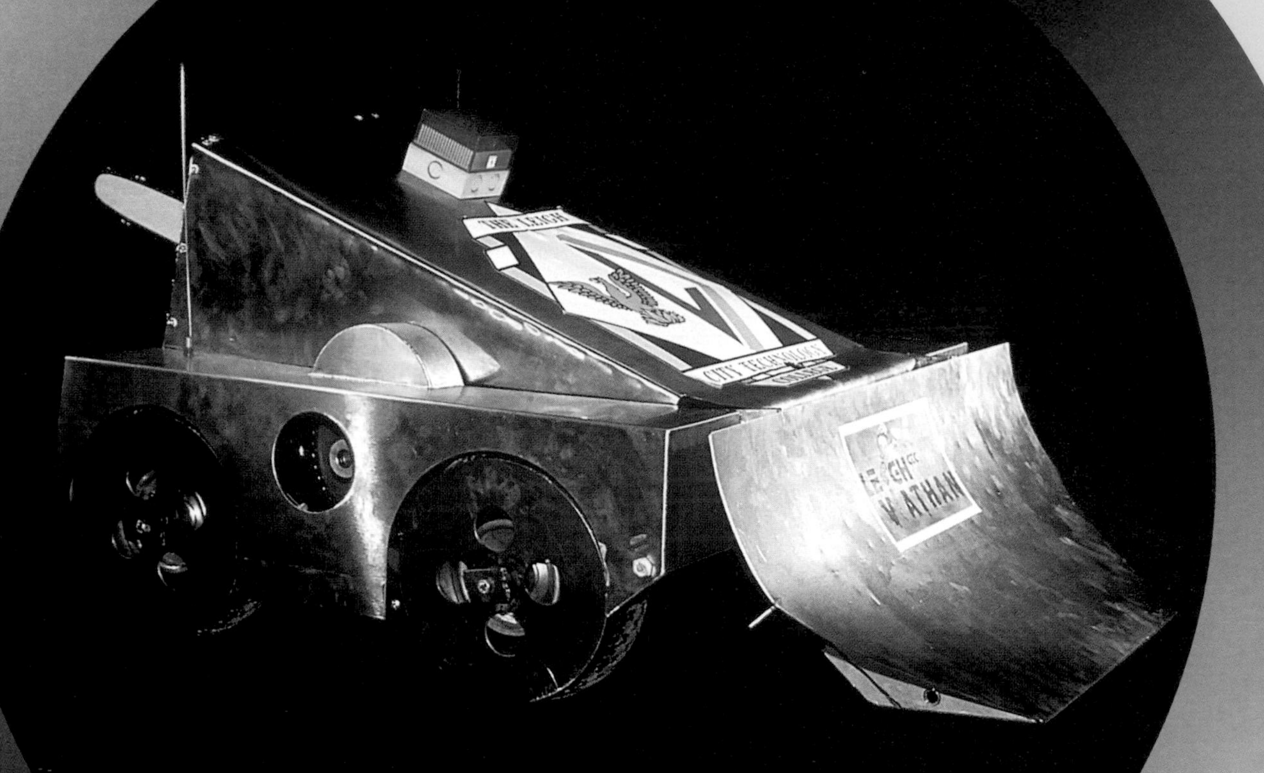

CAPTAIN: ROBIN HERRICK

ROBOTEERS: TONY SCHOFIELD, CHRIS MAYLON

LENGTH: 140 CM

WIDTH: 80 CM

HEIGHT: 60 CM

WEIGHT: 82.9 KG

MAXIMUM SPEED: 16 KPH (10 MPH)

GROUND CLEARANCE: 30 MM

TURNING CIRCLE: ZERO

POWER: 2 X 12-V BATTERIES, RUNNING ON 24V, 2 X 24-V WHEELCHAIR MOTORS WITH 4-WHEEL DRIVE

WEAPONRY: REAR-MOUNTED 40-CM CHAINSAW; FRONT-MOUNTED SCOOP; LASER-GUIDED DIRECTION FINDER

drive than Robot the Bruce, due to its very high manoeuvrability. Nevertheless, George and team-mate Mike Cutter had little trouble during the actual shooting of *Robot Wars*, so reliable and strong was Chaos. The main advantage of Chaos's design is the high torque, which gives it a lot of pulling power, especially in events such as Tug of War. Its main weakness, according to George, is the two-wheel drive and steering, which makes the robot harder to control.

In spite of the various difficulties it suffered, Spin Doctor, built by cousins Peter and Philip Duncanson and Martin Griffin, has several strengths. It is fast, with a maximum speed of 27 kph (17 mph); it is also highly manoeuvrable, and can spin fast on the spot (a feature designed as a victory salute). In addition, the two large, integrated wedge-like features on each side of the robot make it very difficult to grab.

Leighviathan's main strength is its high manoeuvrability, based on a five-wheel system which enables it to spin on the spot. The fifth wheel can also be raised and lowered for increased traction, a particularly useful attribute for events such as the Tug of War.

With hindsight, Tony Schofield of the Leighviathan team thinks that it would have been a good idea to build a lighter-weight chassis and heavier outer armour, illustrating the importance of achieving a workable compromise between the underlying stability of a robot and its structural durability.

Tantrum's main strength is its steel armour plating, a factor that contributed to its excellent time running the Gauntlet. While Tantrum's main weapon, a front-mounted spiked flail made from 25-millimetre diameter sharpened steel, powered by a 300-watt electric motor and running at around 1000 rpm, was an interesting design, Captain Rupert Weeks thinks that it could have been developed further, for greater effectiveness. He also believes a petrol engine for the main drive would have improved the robot's power.

Milly-Ann-Bug, designed and built by Geoff Warren, Martin Dawson, and Ben Weaver, boasts weaponry consisting of a rear-mounted weighted spike, fashioned from a 10-millimetre hardened silver steel shaft, sharpened to a point, and set on rollers to slide out at opponents; and a front-mounted circular saw blade cut in half, and driven by an electric motor in a 'gnaw-and-nibble' scissor action. Milly-Ann-Bug's main strength is that each of its segments is articulated along its central axis, which gives excellent traction and stability, especially when under direct attack. For instance, when one section is lifted up, the remaining section provides a weight and stability that single-section robots lack.

Initial problems with Killertron's speed controller were solved by opting for a 4QD proportional controller, which allows variable acceleration. The main weapon, a pickaxe, is the same as in the first Killertron, with the addition of a lighter flywheel to counterbalance the axe. The reason for this was to allow the use of more batteries for greater power. According to Abdul Degia, Killertron's main strength is the pickaxe, which serves well in defence as well as in attack.

ROCS (Radio Operated Combat System) was built around the chassis of a second-hand wheelchair. It has a two-wheel drive system, with rear-wheel steering, similar to a forklift truck, and an aluminium shell. ROCS's main weapon is a front-mounted lifting arm, converted from a car jack, powered by two electric motors and capable of lifting at least 75 kilograms. Colin Sievers believes that the robot's main weakness was its angular shape (the more angles a robot has, the more opportunities for an opponent to grab it and do serious damage). The team's next machine will have more sloping sides, for greater defence.

The Plunderbird team were very satisfied with the performance of Plunderbird II's weapons, which include a pneumatic steel ramming arm, and also a front-mounted raising blade, similar to a snow plough. When asked if there were any features that it would have been a good idea to include with the benefit of hindsight, Mike Onslow replied that he likes weapons that are visually exciting, such as chainsaws. The team also considered titanium for the outer body shell, but then decided not to use it since, once made, it cannot be reworked or modified.

Ironing out all the problems during design and construction can also, if luck is on one's side, help a team to avoid problems during the actual *Robot Wars* themselves. This was the case with Panic Attack, which proved to be very strong and durable during shooting.

Robo Doc attempts to avoid the spiked pendulum.

The robot's main advantages were strength, manoeuvrability, and power, not to mention the rather stylish paint job with its spider motif. Panic Attack's weapons consist of a set of electrically operated lifting spikes, mounted on the front to ram and lift opponents. Made from hardened steel, the spikes are 30 centimetres long and are operated by a linear actuator which can lift approximately 1 tonne.

After consulting the many US *Robot Wars* Internet sites, the King Buxton team decided to follow a defensive rather than offensive strategy, with the aim of staying in the contest for as long as possible. However, such a strategy is not quite as easy as it might seem, since the robot must take advantage of the greatest power possible within its weight limit. The reason for this is that, if a machine is going to have minimal weaponry, it should be as difficult to push around as possible, and should be able to push opponents with ease. While possessing the necessary power, King Buxton has the advantage of being able to run upside-down. However, its real innovation is that it is also fitted with dome-shaped hubcaps, a particularly clever yet simple feature, that ensures that if it happens to land on its side, the hubcaps will automatically set it on all four wheels again.

Rob Knight of the Mortis team does not think that chainsaws and disc cutters make particularly good weapons. Although they look good on television, says Rob, they can easily destroy themselves if they get caught in their target. Rob was more than satisfied with Mortis's weapon, a 60-centimetre titanium shaft, tipped with a forged steel Japanese Samurai tanto sword-blade, specifically designed to cut through armour. The shaft is capable of turning through 196° five times per second. This speed, in combination with the tanto blade's triangular cross section which splits metal wider the deeper it penetrates, makes Mortis's weapon particularly fearsome.

COMPETITOR ROBOT

PANIC ATTACK

HEAVYWEIGHT

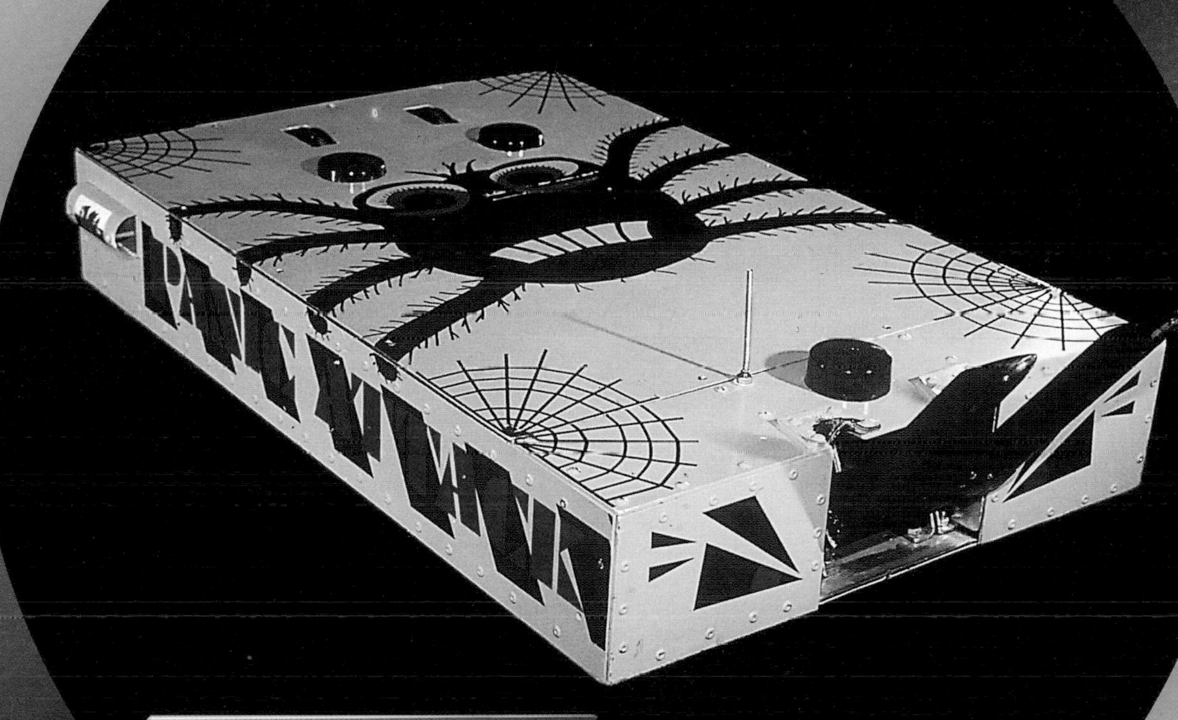

CAPTAIN: KIM DAVIES

ROBOTEERS: KEVIN PRITCHARD, LEE WICOMBE

LENGTH: 125 CM

WIDTH: 70 CM

HEIGHT: 20 CM

WEIGHT: 81.8 KG

MAXIMUM SPEED: 14 KPH (9 MPH)

GROUND CLEARANCE: 25 MM

TURNING CIRCLE: ZERO

POWER: 2 X 12-V (24-AMP) BATTERIES, RUNNING ON 24-V SYSTEM; 2 X 750-WATT MOTORS

WEAPONRY: FRONT-MOUNTED HARDENED STEEL LIFTING SPIKES

COMPETITOR ROBOT

ROADBLOCK

HEAVYWEIGHT

CAPTAIN: HENDER BLEWETT

ROBOTEERS: CHRIS KINSEY, PETER KINSEY

LENGTH: 160 CM

WIDTH: 110 CM

HEIGHT: 45 CM

WEIGHT: 80.9 KG

MAXIMUM SPEED: 8 KPH (5 MPH)

GROUND CLEARANCE: 5 MM

TURNING CIRCLE: ZERO

POWER: 2 X 12-V BATTERIES (30-AMP);
2 X 12-V BATTERIES (10-AMP);
2 X 24-V WHEELCHAIR MOTORS

WEAPONRY: REAR-MOUNTED 30-CM CIRCULAR SAW
WITH TUNGSTEN-TIPPED TEETH

OBTAINING THE MATERIALS FOR CONSTRUCTION

Perhaps the greatest freedom you will experience during your involvement with *Robot Wars* is in actually choosing the materials from which your fighting machine will be fashioned. Indeed, there is no one formula for building a robot, no rule book that says, 'If you do this, this and this, you will end up with the perfect, invincible robot.' In fact, you would be surprised at some of the pieces of apparent junk that have been used in robot construction, and you may consider building a robot to be a good way of clearing away some of the bits and pieces that have been gathering dust in your loft for years!

Hender Blewett heard about *Robot Wars* while the programme was in preparation, and filled out an application form with the intention of building a robot for his A-level Technology project. He went on to become the Champion of the First Wars with his robot Roadblock, and got a grade A for his project into the bargain.

Initially, Hender sat down with his teacher Peter Kinsey (who is team-mate Chris's father) and looked at the robots from the American Wars, taking the best ideas from each one. As is so often the case, money was a big problem. Fortunately for Hender a number of local companies donated materials. The team also wrote to a wheelchair company, who kindly donated a second-hand wheelchair. Roadblock was built around the wheelchair chassis, with its main weapon (a 30-centimetre diameter circular saw blade with tungsten-tipped teeth) powered by the motor from an electric grass trimmer.

Initially, while they had plenty of ideas for how their machine would actually work, the team had problems coming up with a body design. By chance, Hender saw a road sign that exactly fitted the dimensions of the chassis, and which he could see would make an excellent bonnet. Fortunately, the team knew someone who worked at the local council, and this contact took them to a huge skip, from where they chose the Road Ahead Closed sign that became the robot's most distinctive feature.

Having won the First Wars, the team decided to enter Roadblock in the Second Wars, with a new chassis from a more up-to-date wheelchair, and a new Road Ahead Closed sign. The rest of the bodywork is the same as the original Roadblock, since Hender thought the battle damage looked good.

Roadblock tackles the Skittles.

WARS STORIES

Having looked in detail at the various elements that it is important to consider when building your own robot, it is worthwhile learning from the stories of those robots that have competed successfully and unsuccessfully in the First and Second *Robot Wars*. Remember, there is no shame in defeat, but if you are armed with knowledge of what has gone before, you will be better prepared for the future.

Every element of Sting's design was geared to provide a low centre of gravity. This is a very useful feature, since it enabled Sting to maintain a very low ground clearance, and thus deflect attacks from wedge-shaped robots attempting to overturn it. However, the added feature of an adjustable ride height meant that when faced with an obstacle such as a ramp, the running gear could be extended to increase its ground clearance.

For Mark Symons's team, the design of their robot Broot was based on three criteria: a machine that could function upside-down (thus eliminating the danger of overturning); high manoeuvrability, including a zero turning circle; and low cost. For the motor and batteries, they managed to obtain a second-hand wheelchair, the best and cheapest option, giving them pretty much everything they needed.

The chassis was constructed of angle iron, which is made of inexpensive but very tough mild steel. The wheels were made of plywood, and the tyres were made of pieces of water hose hammered on to the wheels with roofing nails. (It should be clear by now that Broot is not a high-tech machine!) In fact, the angle iron in the chassis proved very effective against house robot Matilda's chainsaw.

In spite of its unglamorous appearance and characteristics, Broot is actually a triumph of economy and ingenuity; in fact, when asked if there were any improvements he would have liked to make to the robot, Mark replied that there were none that could have been made, given cost considerations.

Mike Rickard used the flipping mechanism concept in the design of his team's robot, GBH, a rather unusual machine with a highly contoured outer shell, at the front of which is mounted a lifting ramp which is capable of throwing a 76-kilogram weight – as Mike discovered when he stood on it during testing and was thrown into the air! The chassis, based on that of the team's first robot, Scrapper, is constructed of modular square box-section steel, which is very strong but quick and easy to work with. Unusually, the team also used a gearbox instead of speed controllers. The gears have a ratio of 44:1, which gives enormous torque: GBH is capable of pulling a Mini Metro.

Bodyhammer's highly unusual truncated cone design was a refinement of the splendid robot which reached the grand final in the First Wars. The main attributes that the team members Robin Herrick, Andrew Dayton-Lovett, and David Gribble looked for were a strong outer shell, speed, and manoeuvrability. The new robot followed the same basic design as their first one, but featured new weapons: a front-mounted, 15-centimetre diameter circular saw, and a flipper, part of which was salvaged from a Rotavator. Due to the similarities with the first model, the team experienced no particular problems in converting the design to a working robot. Also, the major design parameters were worked out on a computer before construction began.

The corners of George Francis's and Mike Cutter's robot Chaos caused some problems during close-quarters battles: in particular, opponents had a habit of getting stuck on the corners of Chaos when it rammed them. However, as a result Chaos ended up just pushing them around in a circle rather than doing any serious damage. George himself believes that this was more due to a misguided driving strategy than any design flaw. What he should have done when they got tangled, he says, was back up and ram them head on.

Peter Duncanson's involvement with *Robot Wars* came about as a result of watching the First Wars and shouting at his television. There were so many things he would have done differently from the people he saw, he

COMPETITOR ROBOT

BODY HAMMER

HEAVYWEIGHT

CAPTAIN: ROBIN HERRICK

ROBOTEER: ANDREW DAYTON-LOVETT, DAVID GRIBBLE

DIAMETER: 80 CM

HEIGHT: 65 CM

WEIGHT: 77.0 KG

MAXIMUM SPEED: 24 KPH (15 MPH)

GROUND CLEARANCE: 5 MM

TURNING CIRCLE: ZERO

POWER: 1 X 24-V BATTERY; 2-WHEEL DRIVE, WITH 24-V BOSCH MOTORS

WEAPONRY: FRONT-MOUNTED 15-CM CIRCULAR SAW; FRONT-MOUNTED LIFTING ARM

COMPETITOR ROBOT

PIECE DE RESISTANCE

HEAVYWEIGHT

CAPTAIN: COLIN SCOTT

ROBOTEERS: BRYAN NEWCOMBE, JULIE SCOTT

LENGTH: 125 CM

WIDTH: 75 CM

HEIGHT: 65 CM

WEIGHT: 57.0 KG

MAXIMUM SPEED: 8 KPH (5 MPH)

GROUND CLEARANCE 12 MM

TURNING CIRCLE: 120 CM

POWER: 2 X 12-V BATTERIES; 5 X 12-V MOTORS, WHICH ORIGINALLY POWERED CAR WINDSCREEN WIPERS

WEAPONRY: FRONT-MOUNTED STEEL LIFTING SHOVEL; 5 X 2.5-CM REAR-MOUNTED CHROME-PLATED ALUMINIUM SPIKES

decided that he simply had to have a go himself. His team's robot, Spin Doctor, came about more as a result of accident than design. Originally they had wanted a rounded robot, but the availability of materials dictated a much more angular, diamond-shaped construction.

Peter says that everything that could go wrong with conversion from design to working model, went wrong. Every idea the team came up with proved unfeasible, and eventually they settled for a power unit consisting of two Yamaha starter motors from motorbikes, a 30-millimetre box section steel chassis, with an outer shell consisting of polypropylene coated with zinc. The robot took four months to build.

Once they had been selected for the Second Wars, the team discovered that Spin Doctor was the largest robot in the competition. SpinDdoctor then proceeded to get stuck in various places on various occasions. The reason for the oversized shell was that the original motors were not powerful enough, and had to be replaced with the much smaller and more powerful starter motors. This left a great deal of empty space inside the shell, and Peter explained ruefully that the team could have put a lot more equipment in there (such as weapons systems), or they could have made the body much smaller, and not got stuck so much. As is so often the case, however, time was against them.

The essential components – chassis and motors – of the robot Haardvark came from a scooter and wheelchair. But Captain Mike Evans built his own proportional speed controller, which allowed a variable acceleration. The central chassis was strengthened with strip steel and aluminium beams, and was strong enough to carry a 114-kilogram person. In fact, Hardvark, designed and built by Owen Barwick, Mike Evans, and Elizabeth Harrison, boasts a double chassis (the second built around the first), which is enormously strong and resilient. Added to this are two layers of aluminium plating (three around the rear of the robot), which makes the whole machine flame-proof.

For brother and sister Colin and Julie Scott and Bryan Newcombe, the design of Piece de Resistance was something of a hit-and-miss affair; the raw material for the chassis was basically a microwave oven with a wheel mounted on each corner. The outer shell was constructed of fibreglass and wood, and the wheels were taken from two toy tricycles. According to Colin, the robot's main strength is the ease with which it can move forwards and backwards, while its main weakness is that

Sir Killalot's barbecue: Piece de Resistance meets its destiny. Even Matilda can't bear to watch.

it has a large turning circle, is very difficult to steer, and would have benefited greatly from a car-style rack-and-pinion steering system.

Inspiration for a robot's design can come from many places, some more bizarre than others. The Killertron team, consisting of father and son Abdul and Ian Degia, and Richard Broad, found theirs from the lid of a wheely bin, which they used as the foundation of their machine. The design of Killertron for the Second Wars is virtually the same as that for the First Wars, with one or two minor variations, such as orange paint for the body work (in the First Wars, Killertron had a pink colour scheme because Richard bought the wrong paint!).

According to Dave Clutterbuck, the design for Demon was based around cost, feasibility, and a conveniently obtained power source. The team opted for a steel chassis, powered by two wheelchair motors, with a hardened aluminium shell. The weaponry, a set of 10-millimetre steel spikes on the back, is a mixture of offensive and defensive, making Demon difficult to attack successfully from the rear.

For John Reid, the most important design characteristic in Killerhurtz was aggression, which is reflected in

Only the strongest can survive the Gauntlet.

COMPETITOR ROBOT

KILLERTRON

HEAVYWEIGHT

CAPTAIN: RICHARD BROAD

ROBOTEERS: ABDUL DEGIA, IAN DEGIA

LENGTH: 145 CM

WIDTH: 70 CM

HEIGHT: 60 CM

WEIGHT: 75.0 KG

MAXIMUM SPEED: 24 KPH (15 MPH)

GROUND CLEARANCE: 100 MM

TURNING CIRCLE: ZERO

POWER: 2 X 12-V BATTERIES, RUNNING AT 24V;
2 X WHEELCHAIR MOTORS

WEAPONRY: FRONT-MOUNTED PICKAXE

COMPETITOR ROBOT

KILLERHURTZ

HEAVYWEIGHT

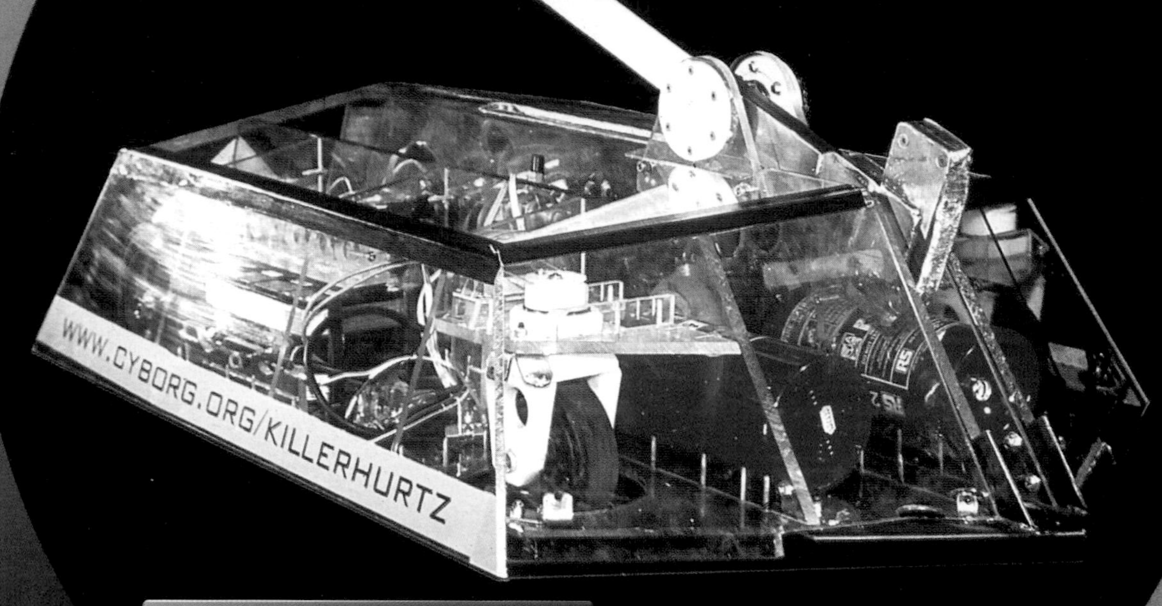

CAPTAIN: JOHN REID

ROBOTEERS: DOMINIC PARKINSON, REBECCA REASTON-BROWN

LENGTH: 115 CM

WIDTH: 80 CM

HEIGHT: 40 CM

WEIGHT: 77.8 KG

MAXIMUM SPEED: 24 KPH (15 MPH)

GROUND CLEARANCE: 10 MM

TURNING CIRCLE: 1 METRE

POWER: 8 X 24-V CYCLON BATTERY CELLS;
2 X 24-V BOSCH MOTORS

WEAPONRY: FRONT-MOUNTED PNEUMATIC STEEL AXE

The Killerhurtz team carry out repairs on their deadly machine.

the robot's appearance. The sleek, transparent shell, through which can be seen the complex internal mechanics and circuitry, is dominated by the weapon, a 3-kilogram custom-built steel axe powered by a pneumatic cylinder generating 800-kilograms of force. John races model cars, so he opted for a similar steering method, controlled by two independent servos.

John says that the main problem for the Killerhurtz team was money, although in some ways this was a blessing in disguise, since it helped them to be highly disciplined in their thinking about what materials to use. To save them the time, money and the trouble of building unsuccessful prototypes, the team designed Killerhurtz on a computer. They decided to use a polycarbonate sheet for the shell, since it is extremely tough and resistant to impact.

Apart from this, Killerhurtz's other main strength is its 24 kph (15 mph) speed, and the power of its 96 kph (60 mph) axehead. Its main weakness is that the polycarbonate shell is not particularly resistant to circular saws. As a result of their experiences when building Killerhurtz the team were amazed at the wide variety of robot designs they encountered during the Wars.

Plunderbird II, designed and built by Mike Onslow, Bryan Kilburn, and Ken Burt, is a refinement of the gorgeous and much-loved Plunderbird I, which won the Best Design Award at the end of the First Wars. Although Plunderbird I looked wonderful, it wasn't particularly well engineered. Mike Onslow designed the robot's tracks, but perhaps because he is a sculptor and mouldmaker, the tracks did not work very well.

Plunderbird II is much more powerful and better engineered than its predecessor. It is powered by a Parvalux industrial motor, and during the tests, Plunderbird II managed to pull Ken Burt's Range Rover and a Ford Transit mini bus. This time, the tracks were designed and built to the team's specifications by Mike's

The Plunderbird team: ready for action.

friends, John and Fiona, a couple living across the street from him, who are both engineers. In addition, independent suspension was added to the wheels.

The robot was designed on paper prior to work beginning on construction. The outer shell was constructed by folding a single sheet of steel into shape. This has the advantage of giving enormous strength and durability to the overall machine. The degree of strength is readily attested to by the fact that Plunderbird II was the only robot to defeat both Dead Metal and Sergeant Bash inside a minute.

The Plunderbird team have established a reputation as very colourful characters; they have even composed a song for each of the First and Second Wars: 'The Plunderbird Song' and 'Robot's Revenge', respectively.

Despite the very successful refinements made to the basic Plunderbird design, the new machine remains decidedly low-tech, but robust. The biggest problem encountered by the team during the Second Wars was the danger posed by flame-throwers, although they managed to solve this by covering the weapons bay and grills along the top of the robot with aluminium tape.

After seeing robot Biohazard in the US *Robot Wars*, Kim Davies of the Panic Attack team decided to go for a similar design, with a flat, low profile. Kim says that the team were beset by a huge number of problems during

COMPETITOR ROBOT

PLUNDERBIRD II

HEAVYWEIGHT

CAPTAIN: MIKE ONSLOW

ROBOTEERS: BRYAN KILBURN, KEN BURT

LENGTH: 90 CM

WIDTH: 110 CM

HEIGHT: 37 CM

WEIGHT: 81.3 KG

MAXIMUM SPEED: 16 KPH (10 MPH)

GROUND CLEARANCE: 20 MM

TURNING CIRCLE: ZERO

POWER: 2 X 12-V BATTERIES, 1 FOR LOCOMOTION, 1 FOR WEAPON; PARVALUX INDUSTRIAL MOTOR

WEAPONRY: FRONT-MOUNTED PNEUMATIC STEEL RAM; FRONT-MOUNTED RAISING BLADE

the design and construction process. For instance, it took four rebuilds to weld the sprockets to the drive wheels correctly. Eventually, though, they managed it, and the four-wheel drive system allows a zero turning circle, which dramatically increased manoeuvrability. In addition, the initial gearing system proved too powerful and hard to control, so the team geared it down, sacrificing speed for greater ease of control. This illustrates how a compromise, which might seem unattractive at first, can result in a better-designed robot.

At first sight, Corporal Punishment looks like a rather run-of-the-mill box on wheels, albeit with an interesting weapon mounted on the front. However, this humdrum exterior hides a highly original internal design. The internal components are suspended independently of the outer shell, resting on a high-pressure nylon hose. There are also spiral steel hawser springs mounted on the inside. The purpose of this elaborate arrangement is to cushion the robot's internal mechanisms from external shocks.

Mortis was a heat winner in the First Wars. The initial team, comprising Rob Knight and Chris Sorsby, watched the US *Robot Wars*, and also consulted the various sites on the Internet, before coming up with the design for their machine. They decided to use tracks, which turned out to be the most vulnerable part of the robot due to the extreme toughness of the outer armour, which was constructed of titanium laminated with kevlar. For the Second Wars the team decided that they needed some moral support and expanded themselves to include Arthur Chilcott and Ben Impey.

Once a house robot gets you, the others are sure to follow.

COMPETITOR ROBOT

MORTIS

HEAVYWEIGHT

CAPTAIN: ROB KNIGHT

ROBOTEERS: ARTHUR CHILCOTT, BEN IMPEY

LENGTH: 130 CM

WIDTH: 75 CM

HEIGHT: 39 CM

WEIGHT: 78.3 KG

SPEED: 14 KPH (9 MPH)

GROUND CLEARANCE: 50 MM

TURNING CIRCLE: ZERO

POWER: 8 X 12-V YUASA HIGH-RATE DISCHARGE BATTERIES; 3 X SIBA COMPOUND CUMULATIVE WOUND ELECTRIC CAR MOTORS, 24-V SYSTEM, BOUGHT FROM SCL MOBILITY; TANK-STYLE TRACKS; RENOLD CONVEYOR CHAIN

WEAPONRY: FRONT-MOUNTED TANTO SWORD BLADE

WEIGHT CLASSIFICATIONS

THERE ARE FOUR CLASSES OF ROBOT:
FEATHERWEIGHT: UP TO 11.4 KILOGRAMS
LIGHTWEIGHT: OVER 11.4 KILOGRAMS TO 22.7 KILOGRAMS
MIDDLEWEIGHT: OVER 22.7 KILOGRAMS TO 45.4 KILOGRAMS
HEAVYWEIGHT: OVER 45.4 KILOGRAMS TO 79.4 KILOGRAMS

There are no restrictions on the size of your robot, although practically speaking machines with the best chance of doing well will probably be approximately 127 centimetres long and 76 centimetres wide. Likewise, your robot can be as tall as you wish, but remember that very tall robots will suffer from instability, and will be vulnerable to overturning by opponents. The maximum voltage for electric motors is 30 volts DC (direct current), or 50 volts AC (alternating current).

Although in the televised Wars emphasis is placed on Heavyweight robots, particularly in the main competition, there is still plenty of scope in the Wars for robots of other weight classifications. Other categories, including a new Super Heavyweight division, are featured. It is well worth considering building a machine for a category other than Heavyweight, particularly if you have little or no experience in robot construction.

For those thinking of building a machine lighter than Heavyweight, it is worth running through the other classifications as well as Heavyweight:

FEATHERWEIGHT

This is a good weight for beginners, who have no prior experience of the sport, and who do not know a great deal about mechanics, electronics, or construction (remember, you *don't* have to be an engineering genius to take part in *Robot Wars* and have a great time!). Many of these robots are based on commercially available radio-controlled model cars, which come in a variety of specifications. However, some of the cheapest kits use the 27MHz AM radio frequency, which is not allowed in *Robot Wars*, due to its susceptibility to interference. Nevertheless, it is possible to change these AM frequencies to 27MHz FM, or 40MHz – the preferred frequency for *Robot Wars*.

If you buy a commercially available kit, you may wish to use the chassis as is, since the receiver and batteries will already be in their optimum positions,

with little or no need for modifications. On the other hand, you may wish to install the drive into a chassis of your own design, which has the advantage of allowing space for extra batteries. But remember to make sure that the chassis you choose is strong enough to carry the extra battery weight at a respectable speed.

It is not a good idea to use the actual bodywork of the kit you have bought, for two reasons. Firstly, such bodywork is likely to be utterly destroyed in the first few minutes of action in *Robot Wars* (imagine a plastic-bodied radio-controlled kit car facing Sir Killalot, and you will appreciate the problem); and secondly, such a strategy shows a paucity of imagination and ingenuity that will not be well received either by the crowd or by the Judges.

Alternatively, there is nothing to stop you building the entire robot from scratch, including the chassis. It is entirely up to you.

For lighter robots the battlefield is just as deadly.

LIGHTWEIGHT

Lightweight robots are similar to Featherweights, but have the potential to include larger motors and some limited weaponry, while still working perfectly well with components designed for use in commercially available radio-control kits. This weight category is ideal for those with some experience in building radio-controlled cars.

MIDDLEWEIGHT

At this weight, shop-bought kit components become unfeasible, since they are rarely strong enough to withstand the strain placed on them by the extra weight. Nevertheless, it is worth experimenting, since some drive components designed for radio-controlled racing cars may be powerful enough. The Middleweight category is a good starting point for those with more experience in either engineering or sophisticated modelling, since the extra weight allows for more imaginative design in terms of active features such as weapons.

HEAVYWEIGHT

As previously mentioned, most of the action in *Robot Wars* will feature Heavyweight robots. This is by far the most challenging (not to mention visually exciting) category, and will require much more in the way of engineering and driving skills, design ingenuity, and imagination. If you feel you're up to the challenge (and there's no real reason why you shouldn't be), and are comfortable working with large, powerful motors, sophisticated weaponry and a solid, durable chassis, then why not opt for the drama and excitement of entering a Heavyweight robot?

Once you have listed all the attributes and components you would like in your robot, you may well find that they place it over the maximum weight allowed for its class. This is particularly hazardous for a Heavyweight machine, since its maximum weight is also the maximum weight for the main competition. Many roboteers have encountered this problem, and it is thus instructive to look at the often ingenious solutions they have come up with to squeeze their machines into the weight allowance.

The main problem encountered by the Razer team was that the central weapon, capable of exerting a force of 3 tons at the tip, was extremely heavy, being constructed of three strips of 5-millimetre plate. The outer body was constructed of an aluminium and steel combination, which was so heavy it made the chassis sag in the middle. This problem was solved through the use of a high-tensile bolt through the middle of the robot.

Unfortunately, however, this was not the end of the team's weight problems. Razer was tested in London with no outer skin. The piercing arm itself was only semi-complete, and needed an additional cowling. Razer weighed in at 77.4 kilograms, leaving the team only 2 kilograms for an outer shell which was expected to weigh about 5 kilograms. Simon managed to solve this problem by drilling no fewer than 61 holes in each of the central steel load bearers. These 20-millimetre holes ended up saving the team 1.6 kilograms, which, with the remaining 2 kilograms, was just enough for the outer shell.

Captain Shane Howard of the Mace team watched the First Wars with great interest, paying special attention to the robots' performances. He then tried to design a machine that could, potentially at least, successfully combat all of them. Like many roboteers, Shane and team-mate Brian Fountain discovered that the feature they would most like to have included placed their machine over the maximum weight allowed. Originally, they had intended to include a 5-millimetre steel plate, which would have extended across the entire front of the robot. But in the end they had to settle for a single, variable height ram/lifting arm on the front (capable of lifting 114 kilograms), and a high-speed flail with four blades running at 2000 rpm at the rear.

According to Rex Garrod of the Cassius team, weight is a robot's number-one weapon. Despite this, the team had to plane off 1.5 kilograms of rubber from Cassius's solid go-kart wheels to bring its weight within the specified allowance.

Mike Rickard says that he wanted to include a mechanism by which GBH's ride height could be adjusted; unfortunately, once again such a feature would have put

A wedge faces a bulldozer in a Super-Heavyweight battle to the death.

the robot over the weight limit. Indeed, GBH was originally 12 kilograms overweight, and the team had to install lighter batteries, in addition to replacing a steel bar, which acted as a hinge for the ram, with an aluminium bar.

Leighviathan was designed and built by students at the Leigh City Technology Centre (which entered Leighbot in the First Wars), and was based on a miniature JCB. The JCB-inspired design required the inclusion of a lot of motors and other mechanisms, which increased the robot's weight dramatically, and proved a frustrating and ongoing problem for the team during construction. Initially, they had wanted to include a moving front-mounted bucket/scoop, although in the end they had to settle for a fixed one. The construction proceeded with a certain haphazard charm from an initial thumbnail sketch, to a prototype that didn't work, through various improvements, to a working model. In addition, during construction the team connected a battery the wrong way around, and blew the radio receivers.

Robo Doc team captain Mike Franklin felt that the main problem his team encountered during the design process was weight. For this reason, they decided to go for an aluminium construction. Also, in the original design, the main weapon was a chainsaw capable of pivoting through 180° (a little like Dead Metal's circular saw arm). However, weight considerations again made this unfeasible, so instead the team opted for a lightweight (3-kilogram) lifting arm, which is capable of lifting 70 kilograms.

Adam Clark had problems keeping down the weight of his robot, Corporal Punishment, the reason being that he neglected to include bolts, washers, brackets and so on in his weight calculations – a salutary lesson for all prospective roboteers. Remember to count *every single component* of your machine during these initial weight calculations.

Corporal Punishment's main strengths are its high speed and acceleration, robustness, and the multi-pronged ram/lifting scoop, which has a lifting force of 113 kilograms The main weakness is that the 10-millimetre tube from which the scoop is constructed is not strong enough, and can bend, reducing the robot's ground clearance.

DEALING WITH THE HOUSE ROBOTS

Barring accidents, which do happen, the house robots work in combination with each other. And while they set the standards in roboteering, they don't know what to expect from the competitors' robots. They all have their strengths and weaknesses. While most teams agreed that all the house robots were fierce enemies and equally difficult to deal with, here are some examples of the actions in which they were each involved during the Second Wars, as told by the competitors.

SHUNT

While none of the competitors cited Shunt as their worst enemy, he caused untold damage in combination with the other house machines. Despite this the Spin Doctor team thought Shunt was the easiest house robot to deal with, because its own shell was highly resilient to impact and on one occasion, Shunt was clear to attack him, but bounced off ineffectively. Of course, Shunt is keen to avenge this embarrassing incident. Milly-Ann-Bug was also pleased with her encounter with Shunt as her kevlar coating rendered his axe ineffective.

MATILDA

For the Spin Doctor team, Matilda presented the greatest threat, since her chainsaw proved the only weapon capable of cutting through Spin Doctor's polypropylene shell.

DEAD METAL

This was the Napalm team's favourite house robot, particularly in terms of its design. For the Sting team, the only threat from the house robots came from Dead Metal, which grabbed Sting during the Gauntlet, but luckily did little damage due to the great strength of the robot's armour. However, the team all agree that this was the most heart-stopping moment from the Wars.

SERGEANT BASH

According to David Crosby of the Napalm team, the main threat from the house robots was Sergeant Bash's flame-thrower, which enclosed their machine in fire at one point during the competition. This can prove fatal to a robot, since it is very easy for the internal circuitry to be burned to a crisp. Even if this does not happen, it only requires a single wire to be damaged for a robot to be 'killed'. This is where titanium came in handy as a protective covering for Napalm's wiring. While Neil Lambeth and John Ebdon's Elvis did not have any real problems with the house robots, the front-mounted lifting platform and rear-mounted chainsaw giving adequate protection, the only serious danger was presented by Sergeant Bash's flame-thrower, which fried Elvis's radio receiver.

Peter Gibson considered all the house robots to be equally dangerous, although in contrast to the others he had a particularly successful encounter with Sergeant Bash, when Wheelosaurus's spiked tail managed to puncture the gas cylinder powering the Sergeant's flame-thrower, putting it out of commission. Nevertheless, the spiked tail itself gave problems, when Peter tried a wheely on the wooden floor of the Arena, and a spike got stuck. He later confided that it would have been a good idea to include a limit stop to the tail.

According to Ozzie Boulter, the most dangerous house robot was Sergeant Bash. At one point, the Sergeant melted Wizard's plastic hat, although Ozzie didn't mind, because it looked good, and didn't inhibit the functioning of the robot. This is actually something of a tribute to Ozzie's sportsmanship and sense of fun; in fact, fun was his main concern in *Robot Wars*, and his attitude is an important reminder that maintaining one's sense of humour is essential in this sport.

The Milly-Ann-Bug team also had problems with Sergeant Bash and his flame thrower – problems which were made all the worse by its own main weakness – the speed controller – which suffered seriously whenever its wheels were prevented from turning. The main reason for this was its outer coating of kevlar which, though tough, was not very effective against sustained heat.

SIR KILLALOT

Ian Lewis and Simon Scott of the Razer team had no real problems with the house robots, except for Sir Killalot, the first sight of which gave them a nasty surprise, and they immediately decided to avoid him at all costs! It would seem that this is probably the best strategy for dealing with the newest recruit to the house robots. The Mace team also found the Sir Killalot to be the most fearsome house robot, not least because of his powerful hydraulic pincers, which cut through Mace's chassis. Oliver Steeples too was in awe of the dreadful Sir Killalot, although Griffon did manage to sever one of his hydraulic cables.

For the Chaos team Sir Killalot was far too heavy and too well-armed to deal with. For Abdul Degia and the Killertron team, Sir Killalot seemed more like a tank than a robot; although as far as the actual threat it posed, Abdul felt that it was more or less equal with the other house robots – a view not shared by all the other competitors.

Mike Onslow of the Plunderbird team (who successfully dealt with both Dead Metal and Sergeant Bash) said Sir Killalot was 'awesome'. Nevertheless, Mike outlined a possible strategy for defeating the beast, which involves a wedge-shaped robot reversing up to it, and then a smaller robot riding up the wedge at full speed to smash the house robot in the face. An interesting idea, although it may prove difficult to find volunteers!

HINDSIGHT

One of the most interesting aspects of conversation with roboteers is discovering how they *would* have designed their machines, with the benefit of hindsight. Far from being a futile exercise, this can point the way to essential refinements and improvements in design that may give a subsequent robot the edge in future battles, in addition to illustrating the many pitfalls awaiting the novice roboteer.

David Crosby, who built Napalm, Ian Lewis of the Razer team, and George Francis who built Chaos, all believe that it would have been a good idea to include a self-righting feature in their designs, since, once you have been overturned by an opponent, the chances are you've had it. Nevertheless, Napalm's weight distribution gives it a very low centre of gravity, which in turn makes it extremely difficult to overturn. According to David, another excellent weapon would have been some kind of climbing rope by which the robot could pull itself up out of the Flame Pit. In addition Ian thought that a rear-mounted grinder would also have been useful, but these options would have put Razer well over the maximum weight allowed for a Heavyweight robot. Another good idea, he said, would have been to save weight in the chassis by using titanium instead of steel for the load bearers.

In contrast to Ian and David, however, when asked about the inclusion of a self-righting mechanism, Shane Howard of the Mace team, thought that if you need such a mechanism, you're probably not such a good driver, and you shouldn't be in *Robot Wars* in the first place!

Neil Lambeth of the Elvis team believes that it would have been a good idea to include a more powerful motor, and a stronger lifting mechanism. When deployed Elvis's lifting mechanism did not really have enough power to cause serious problems to his opponents.

Shane Howard of the Mace team believes that more powerful drive motors would have been a good idea, for the simple reason that one of the best weapons a robot can possess is the power to push an opponent into

Mortis takes a chunk out of the hapless Oblivion.

obstacles or traps. In addition, he says that Mace's lifting arm should have worked much faster than it did.

On the whole, Oliver and Ben Steeples were pleased with their design for Griffon; the only element they would have added was a six-wheel drive, for greater speed and traction.

According to David Barker, Sting could have been made even stronger by the inclusion of heat shielding against a flame-thrower, such as the one in the Flame Pit which he encountered during the Gauntlet and in the Arena. A robot's more delicate components require such protection, and it was only through lack of time that the team did not install it in Sting.

So successful was Cassius's design that it managed to overturn all the house robots except Sergeant Bash. However, team captain Rex Garrod was not entirely satisfied with the design of the outer shell, which he felt was rather too large, and left a good deal of unused space inside. Cassius, he says, could also have done with a stronger shell. Rex would have preferred a UHF control system, since the receiver mast only has to be about two inches long, and so would be much less vulnerable to the attentions of opponents' robots. He also thinks that a variable ride height is an excellent idea, since it enables a robot to deal with a variety of potential threats, with low ground clearance to avoid being overturned by wedges, and higher ground clearance to negotiate obstacles such as ramps.

Ozzie Boulter believes that it would have been a good idea to use larger and wider wheels on his robot Wizard, without tread to improve grip on dry surfaces (Wizard's wheels were just over 6 centimetres in diameter and just over 1 centimetre wide).

Kim Davies of the Panic Attack team felt that his robot's flat sides had been a bad idea. In future, he has decided that they would go for more sloped sides to deflect attacks. Kim believes the most challenging aspect of the Wars themselves is the Sumo round, which demands extreme control of one's robot.

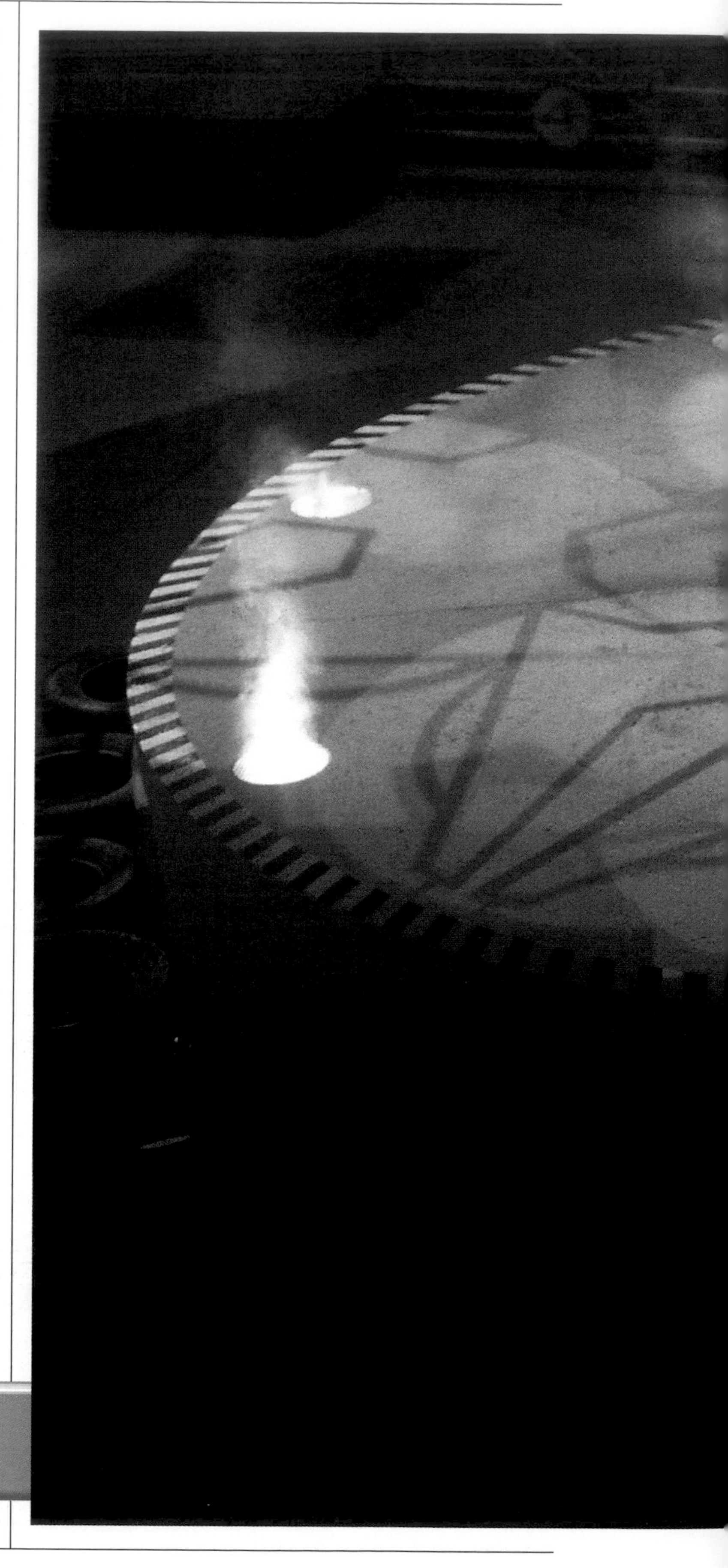

Panic Attack playing Sumo: harder than it looks, with Shunt as your opponent.

ROBOT WARS ON THE INTERNET

Not surprisingly for a show that has its roots in technology, *Robot Wars* has invaded the Internet on all fronts. In common with many cult programmes, it is supported, promoted and discussed on-line by the competitors and the fans.

While many unofficial sites exist – often designed by the roboteers themselves – the official website for *Robot Wars* is perhaps the most useful on-line resource available. It gives visitors unprecedented access to the world of *Robot Wars*, behind-the-scenes information and contact with the production team. The site – which emulates the programme's styling – is the perfect place to get advice from other builders and keep up to date on all the latest news and developments. More importantly there is no better way to make new contacts, communicate with other fans and interact with competitors from all over the world.

Roboteers are advised to surf before they build and fans of the programme are urged to become part of the on-line *Robot Wars* community.

When you first enter the site at www.robotwars.co.uk you will find yourself on the homepage where there is a scrolling panel with all the hot-off-the press news about *Robot Wars*. From here you can venture onto a number of other screens:

BRIEFINGS

The latest news, developments and officially leaked rumours from the world of *Robot Wars*. If you hear it anywhere else, it isn't necessarily true.

INTELLIGENCE

An ever-growing archive of information on the *Robot Wars* television show, including details on the presenters, house robots, production team and much, much more.

RECRUITMENT

Here you can contact the production team, join the *Robot Wars* Club or build a killing machine. *Robot Wars* needs you!

COMMUNICATIONS

This is where you can communicate with the people behind *Robot Wars*, fans of the programme and the competitors themselves by posting a message on the bulletin board or by chatting live.

STORES

The only place to buy official *Robot Wars* merchandise and join the *Robot Wars* Club via on-line commerce.

ARCHIVES

The definitive archive of production photographs, artwork, sound and video files. This includes the music from the show, the presenters' 'bloopers' and the conceptual artwork used during the show's development.

LINKS

Containing links to many unofficial sites connected to the world of *Robot Wars* and robotics in general.

THE ROBOT WARS CLUB

So, now that you've read through this book how do you feel about building your own robot? Are you ready to wrestle with the problems described here; to experience the oil, sweat and tears, the agony of an overheated speed controller, or the frustration of a lifting arm that couldn't overturn a packet of crisps? Are you ready to march steadily through adversity, with your sights firmly set on the satisfaction of winning a heat, and the greater glory of becoming a series champion? Or perhaps the excitement of competing in front of thousands of robot-mad spectators during the *Robot Wars* live events inspires you?

If so, then there is one thing it is absolutely essential that you do: you must join the *Robot Wars* Club, for without club membership, you will not be able to enter a machine in the Wars. Membership of the *Robot Wars* Club costs £10, for which you will receive the following:

- A free *Robot Wars* Video of the 1996 American *Robot Wars* final.
- The full-colour *Robot Wars* Club magazine.
- A free first edition enamelled *Robot Wars* metal badge, featuring the series logo.
- Unique *Robot Wars* merchandise offers, exclusive to Club Members.
- Your own personal membership card entitling you to great Club discounts and the opportunity to attend the *Robot Wars* Championship in the US.
- The opportunity to apply for free tickets to the filming of future series of *Robot Wars* in Britain.
- In addition, you will also receive a Fact File, covering the origin of the sport, plans for the future, and information on what to do if you want to compete. This will include an application form for entry into the next series of *Robot Wars*.

Once you have completed this form, you will be notified if you have been selected for the competition. The reason for this selection process is that there is a truly enormous demand for direct involvement with *Robot Wars*, and unfortunately not everyone who applies will be able to take part. However, there are already plans to organize both live events and regional heats, so that, even if you don't make it to the televised event, you may well still be able to contribute to the Wars in another way.

If you are selected, you will then be sent a full set of Rules and Regulations, detailed guidance on how to build a robot safely, and other necessary paperwork. The safety aspect is paramount, and it is for this reason that you should under no circumstances begin designing and building a robot until your application has been processed and approved, and you have received the relevant paperwork. The building of robots is potentially dangerous, and furthermore, it should be noted that the information contained in this book is intended only to give hints and tips on robot construction: it is a *technical* manual, not an *instruction* manual. For detailed guidance, remember, you must join the *Robot Wars* Club. (Apart from anything else, you would not want to invest your time, effort and money in building a robot, only to discover that you have not been selected for the competition!)

Those club members who go on to build a robot will have their club membership fee refunded, and will become honorary Builder Members of the *Robot Wars* Club.

If you would like to join the *Robot Wars* Club, please send a cheque or postal order for £10 to: Robot Wars Club, London W1E 1AW.

HOW TO MAKE A VIDEO DIARY

If you decide to design and build a robot to participate in *Robot Wars*, you may like to consider making a video diary of your project. This will be extremely useful as an accurate record of the various stages in the process of robot construction, not only for you, but also to the producers of *Robot Wars*, who may decide to include segments of it as background material to the televised events. Although not essential, a video diary will add more depth and human interest to your participation, especially to the audience watching at home who are not directly involved with *Robot Wars*, but who still may be interested to see how a robot is designed and built.

Here are some helpful points to help you plan your video diary

- Please introduce all team members by name (or nickname) and what their job is on the team, for example: 'This is John Smith, more commonly known as Smithie, and he's head of Weapons Research.' If there are any interesting or funny anecdotes or facts that are relevant to the team or *Robot Wars*, please feel free to include them in the video.
- It is important to video all the different stages of your robot's construction, for example: design stage, initial construction of chassis, fitting electronics, engine and other components, and also the aesthetic work that you do to complete your robot. Don't worry if you intend to gather your materials from a scrap yard or other unglamorous location; make sure you get it on video nevertheless.
- Don't worry about trying to record a finished piece: the material you supply will have to be edited anyway, and the more footage you record, the easier it will be for the *Robot Wars* production team to use.
- Remember that while *Robot Wars* is a serious sport, it was also conceived as entertainment, so it is important that you maintain a sense of fun while you are making your video diary. Don't take things too seriously: if you have a good laugh and enjoy yourselves, then so will the people watching the diary. Also, try to record team production meetings, and if there are any differences of opinion, please keep them in the video: this will help the audience to understand better how a robot is conceived, designed, and constructed. But please, no violent scenes or bad language!
- Don't just record the successful parts of your project. If you spend a lot of time designing and building a weapon that, when tested, doesn't work, you should show this as well. It will be all the more interesting for the audience to see how you managed to overcome the problem. It is important to show viewers how much technical skill, ingenuity, and creativity is needed to construct a robot.
- Have lots of fun making your video diary, and if you should think of anything to include in it that is not mentioned in this guide, then include it.
- Finally, it would be very helpful if you could give some indication of what order the footage appears on your tape. For example:

A Team introduction
B Design meetings
C Gathering components
D Team discussions
E Construction
F Testing

INDEX

A
aggression 32
Arena 9, 10, 18, 24–5, 28
  building 23–7
audience 33
autonomous machines 32

B
Barker, David 25, 99, 105, 138
Barker, Matthew 99, 105
Barwick, Owen 119
batteries 83
  liquid electrolyte batteries 83
  Nickel-Cadmium (Ni-Cad) battery 83
  Sealed Lead Acid Gel battery 83
BBC Visual Effects Department (VED) 36, 39, 44, 51, 54, 61, 66–7
Blewett, Hender 115
body construction 70, 104–5
  materials 86, 104–5, 115
  styling 99–102
Boulter, Alwyn 'Ozzie' 97, 102, 135, 138
box-shaped robots 75, 99
Brett, Phil 92
Broad, Richard 97, 120
Bull, Ed 86
Burt, Ken 123

C
caption generator 35
Carsey, Stephen 21, 22, 23, 28, 29, 31, 34, 36
Charles, Craig 18, 29, 30, 31, 34, 36, 67
chassis 75
Chilcott, Arthur 126
Clark, Adam 131
Clark, Matt 34
Clarkson, Jeremy 8, 29
Clutterbuck, Dave 86, 120
Cockerill, Nathan 29
commentator 36
commercially available kits 128, 129, 130
competitor robots
  Barry 50
  Behemoth 76
  Bodyhammer 109, 116, 117
  Broot 32, 116
  Cassius 6, 72, 86, 87, 101, 109, 130, 138
  Chaos 84, 86, 109, 111, 116, 136, 137
  Corporal Punishment 126, 131
  Demolition Demon 72–3
  Demon 9, 86, 120
  Dreadnaut 10, 13
  Elvis 16, 33, 91, 95, 99, 135, 137
  Flirty Skirty 86
  GBH 17, 105, 109, 116, 130–1
  Griffon 23, 90, 91–2, 106, 109, 136, 138
  Groundhog 75
  Haardvark 34, 119
  Inquisitor 70
  Killerhurtz 120, 122, 123
  Killertron 44–5, 97, 107, 111, 120, 121, 136
  King Buxton 92, 93, 112
  Leighviathan 110, 111, 131
  Loco 75
  Mace 106, 108, 130, 136, 137
  Mega Hurts 71
  Milly-Ann-Bug 16, 76, 77, 78, 105, 111, 132, 135
  Mortis 112, 126, 127, 137
  Mule, The 9
  Napalm 22, 27, 36, 47, 72–3, 98, 99, 104, 105, 133, 134, 135, 137
  Nemesis 68–9, 99, 102, 103, 109
  Oblivion 137
  Onslaught 109
  Orac 12
  Panic Attack 20–21, 111–12, 113, 124, 126, 132, 134, 138–9
  Piece de Resistance 37, 118, 119–20
  Plunderbird I and II 106, 111, 123–3, 123–4, 125, 136
  Prometheus 62
  Razer 26, 106, 130, 136, 137
  Recyclopse 101, 109
  Roadblock 72, 100–1, 114, 115
  Robo Doc 78, 79, 85, 92, 97, 112, 131
  Robot the Bruce 86, 111
  ROCS (Radio Operated Combat System) 111
  Rottweiler 102
  Scarab 50
  Spin Doctor 22, 105, 111, 119, 132, 133
  Sting 25, 82, 85–6, 99, 101, 105, 116, 134, 138
  Tantrum 78, 111
  Wheelosaurus 22, 74, 75, 77–8, 135
  Wizard 72, 96, 97, 102, 135, 138
control mechanisms 70
crew 18
Crosby, David 36, 99, 105, 135, 137
Cutter, Mike 86, 111, 116

D
Davies, Kim 124, 126, 138
Davis, Bernie 28
Dawson, Martin 111
Dayton-Lovett, Andrew 116
Dead Metal 8, 16, 52–5, 62, 107, 124, 126, 131, 134
Degia, Abdul 97, 111, 120, 136
Degia, Ian 97, 120
design 7, 69–131
  airfoil designs 75
  body construction 104–5
  box-shaped robots 75, 99
  chassis 75
  cylinder-shaped robots 102
  design compromises 70, 111, 126
  drive 80–1, 83
  electric motors 80
  flying robots 75, 94
  hindsight 137–8
  inspiration 120
  legged machines 75, 78
  materials 86, 104–5, 115
  radio control 94, 97
  robot types 70–8
  rules and regulations 69
  self-righting mechanisms 72, 109, 137
  simplicity of design 78
  size and height 128
  steering 85
  styling 99–102
  transmission 88–92
  weaponry and fighting ability 106, 109, 111–12
  'wedges' 72, 75, 99
  weight classifications 128–31
Dickenson, Eric 30, 31
drive mechanism 70, 78, 80–1, 83
  centre-mounted drive 80
  electric motors 80, 83, 85, 94
  internal combustion engines 83, 85, 94
  motor running speeds 86
  traditional drive position 80
driving skills 7, 16, 22
Duff, Paul 27
Duncanson, Peter 22, 105, 111, 116, 119
Duncanson, Philip 111

E
Ebdon, John 33, 91, 99, 135
electric motors 80, 83, 85, 94
  batteries 83
  series connection circuits 80, 83
  upper voltage limit 80, 128
Evans, Mike 119

F
Faraday Cage 94
fibreglass 104, 105
First Wars 8, 9, 10, 11, 27
flying robots 75, 94
Forrester, Philippa 8, 18, 26, 28, 29, 32, 36
Fountain, Brian 106, 130
Francis, George 86, 109, 111, 116, 137
Franklin, Mike 78, 97, 131
Fullalove, Julian 23, 24, 25, 26

G
Garrod, Rex 86, 101–2, 109, 130, 138
Gauntlet 8, 9, 10–11, 138
gears 88, 116
genetic algorithms 32
Gibson, Peter 22, 75, 77–8, 135
Glenn, Becci 22, 36, 99, 105
graphics 34–5
Greenaway, Claire 22, 36, 99, 105
Gribble, David 116
Griffin, Martin 22, 111
Gutteridge, Tom 21, 23, 36

H
Harper, Adam 21, 30–1
Harrison, Elizabeth 119
Harrison, Simon 92
health and safety 22, 25, 33, 106
Herrick, Robin 109, 116
house robots 7, 9, 10, 11, 18, 36, 37, 39–67, 78, 94, 132–6
Dead Metal 8, 16, 52–5, 62, 107, 124, 126, 131, 134
Matilda 8, 10, 12, 13, 38, 46–51, 63, 92, 99, 104, 119, 133
Sergeant Bash 8, 16, 30, 56–61, 83, 91, 107, 124, 135
Shunt 8, 10, 12, 40–5, 126, 132, 138–9
Sir Killalot 6, 8, 10, 12, 17, 25–6, 27, 37, 62–7, 85, 86, 104, 105, 119, 126, 129, 136
Howard, Shane 106, 130, 137–8
Howarth, Craig 35
hydraulics 66–7, 85, 106

I
Impey, Ben 126
Internet 140

J
Joust 13
Judges 30–2
Dickenson, Eric 30, 31
Harper, Adam 21, 30–1
judging criteria 30, 32
Sharkey, Professor Noel 30, 31–2

K
kevlar 105, 126, 132, 135
Kilburn, Bryan 123
King of the Castle 12
Kinsey, Peter 115
Knight, Rob 112, 126

L
La Machine 8, 75
Lambeth, Neil 33, 91, 99, 135, 137
legged machines 75, 78
Lewis, Ian 26, 106, 136, 137
lighting 23, 25, 28

M
McDonald, Stuart 28, 33
Making of Robot Wars (documentary) 8
manoeuvrability 10, 70, 80, 86, 111, 116, 126
the Master 8
materials 86, 104–5, 115
Matilda 8, 10, 12, 13, 38, 46–51, 63, 92, 99, 104, 119, 133
Maze 10, 22
Mentorn Barraclough Carey 8, 21, 36
Monk, Steve 92

N
Newcombe, Bryan 119
Nicholas, Celia 33

O
Onslow, Mike 111, 123, 136

P
Pearce, Jonathan 18, 36
Peter, Richard 86
Pinball 17
plastic 105
Pritchard, Ian 25, 105
programming actions into a robot 32

R
radio control 94, 97
channels, number of 94
electronic fail safes 94, 97
frequencies 94, 128
interference 94, 97
positioning the receiver 94
proportional channels 94
speed controllers 94
UHF transmitters 97, 138
Reid, John 120, 123
Reynolds, Chris 51, 54, 61
Rickard, Mike 105, 109, 116, 130
robot types 70
Robot Wars: concept 7–9
Robot Wars Club 69, 141
Rott, Dominic 102

S
safety *see* health and safety
Schofield, Tony 111
Scott, Colin 119
Scott, Julie 119
Scott, Simon 26, 106, 136
Second Wars 9, 10, 11, 18, 27
self-righting mechanisms 72, 109, 137
sensing devices 32
Sentinal 32–3
Sergeant Bash 8, 16, 30, 56–61, 83, 91, 107, 124, 135
servos 97
Sharkey, Professor Noel 30, 31–2
Shunt 8, 10, 12, 40–5, 126, 132, 138–9
Sievers, Colin 111
Sir Killalot 6, 8, 10, 12, 17, 25–6, 27, 37, 62–7, 85, 86, 104, 105, 119, 126, 129, 136
Skittles 15, 105
Soccer 16, 91, 106
Sorsby, Chris 126
speed controllers 94
spinning on the spot 80, 88, 111
stability 75
steel 105
Steeples, Ben 23, 91, 138
Steeples, Oliver 23, 91, 106, 109, 136, 138
steering 85, 88
car-type steering 88
microprocessor-controlled 92
rack and pinion 88, 120
tank-type 88
styling 99–102
imaginative flair 70, 99, 101, 102
strength and durability 99, 101
Sumo 12, 44, 138
suspension 88
Symons, Ben 32
Symons, Mark 116

T
tank traps 11
teams
number limits 69
teamwork and co-operation 69
technical problems
burn-out 80
ground clearance 70, 102, 109, 116, 138
leverage problem 106
overheating 66, 78, 86, 92
overturning 80, 109, 137
thermal circuit breakers 83
Thor 8
Thorpe, Marc 7
Tilley, Colin 66, 67
titanium 105, 135, 137
transmission 88–92
gears 88, 116
steering 88
suspension 88
wheels 91–2
Trials 8, 9, 12–17
Joust 13
King of the Castle 12
Pinball 17
Skittles 15
Soccer 16, 91, 106
Sumo 12, 44, 138
Tug of War 14
Tug of War 14, 111

V
variable ride height 138
video diary, making 142

W
Warren, Geoff 78, 111
weaponry and fighting ability 72, 106, 109, 111–12
chainsaws and disc cutters 72, 106, 112
concussive weapons 106
flame-throwers 57, 61, 135
hydraulic weaponry 66–7, 106
interchangeable weapons 106
minimal weaponry 112
prohibited systems 22, 106
rotating weapons 72, 106
visually exciting weapons 111
Weaver, Ben 111
wedge-shaped robots ('wedges') 72, 75, 99
Weeks, Chris 78
Weeks, Rupert 78, 111
weight classifications 9, 128–31
Featherweight robots 128–9
Heavyweight robots 9, 128, 130–1
Lightweight robots 9, 130
Middleweight robots 130
Super Heavyweight robots 128, 131
weight calculations 131
West, Simon 86
wheels 91–2
pneumatic tyres 91
solid wheels 92
wood 104
writing the script 29